Forever Open, Clear and Free

Forever Open, Clear and Free

The Struggle for Chicago's Lakefront

Second Edition

by Lois Wille

The University of Chicago Press
Chicago and London

Frontispiece: Aerial view of the Museum of Science and Industry with public beach, marina, and Lake Shore Drive in the background; Jackson Park, 1985. Detail of photo by Ernst H. Seinwill, courtesy of the Newberry Library.

Originally published in 1972 by Henry Regnery Company, in cooperation with the Metropolitan Housing and Planning Council. This edition published by arrangement with Contemporary Books, Inc.

The University of Chicago Press, Chicago 60637
The University of Chicago Press, Ltd., London

 University of Chicago Press edition 1991
Printed in the United States of America

99 98 97 96 95 94 93 92 91 6 5 4 3 2 1

Library of Congress Cataloging-in-Publication Data

Wille, Lois.
Forever open, clear, and free : the struggle for Chicago's lakefront / by Lois Wille.—2nd ed.
p. cm.
Includes bibliographical references.
ISBN 0-226-89871-7 (cloth).—ISBN 0-226-89872-5 (pbk.)
1. City planning—Illinois—Chicago. 2. Waterfronts—Illinois—Chicago. I. Title.
HT168.C5W55 1991
307.1—dc20 90-25852
CIP

∞ The paper used in this publication meets the minimum requirements of the American National Standard for Information Sciences—Permanence of Paper for Printed Library Materials, ANSI Z39.48-1984

Dedicated to the children of Chicago
in memory of Howard E. Green, civic statesman and former president
of the Metropolitan Housing and Planning Council

Contents

List of Illustrations

Foreword to the Second Edition

Forever Open, Clear and Free is a kind of history grown scarce in recent decades. It is not simply an exposé that provides an unrelenting disclosure of things gone wrong or men gone bad. Nor does it pursue some universal thesis densely argued in an academic vocabulary. It is popular history in the best sense of that word. *Forever Open, Clear and Free* makes available to all literate Chicagoans a selective account of those exemplary events, foul or fair, which have produced Chicago's magnificent lakefront parks. More than any other single feature, these parks, extending for almost twenty miles, distinguish Chicago from the other Great Lakes cities. As Wille tells us, however, this accomplishment was not a simple act of collective resolve but a tortured struggle which Chicagoans must never forget.

The great struggle for Chicago's lakefront is not over. Contemporary attempts to divert the lakefront to commercial or private use are unrelenting. Those, like myself, who would defend or extend this parkland are an embattled minority who often find themselves trapped in a rhetoric that juxtaposes elitism to popular usage or economic development to urban refuge. Wille's story must be made part of Chicago's collective memory if we are to find our way past this kind of verbal manipulation.

Wille tells the story with great force and color. There are fur traders, Indians, robber barons, great magnates, ward heelers, visionaries, geniuses, and peacemakers. There are pitched battles, intricate political maneuvers, and marvels of rhetorical invention. But this is no tour guide for visitors or brochure from the chamber of commerce. Wille's heroes are flawed and her villains are complex. Sometimes they are the same people. Take Long John Wentworth for instance. Wentworth was the political point man for Stephen Douglas's seizure of the South Side lake frontage for the Illinois Central Railroad in 1852. Yet it was Long John who, as mayor, cleared Cap Streeter and his squatters off the Sands and reformed—at least temporarily—the police force. Paradoxically the Illinois Central's breakwater became the outer barrier for the land infill which constitutes the lakeside park on the South Side. Very little of Chicago's lakefront parks actually belongs to the original shoreline. It is virtually a man-made shoreline, rescued from early abuses and encroachments.

Or take Cap Streeter himself. An incorrigible obstructionist, he became something of a folk hero when he turned the accumulating sandbar north of the Chicago River into a shantytown in defiance of the city's leading citizens. Yet "Streeterville" has become one of the city's most sought-after addresses, not least because wealthy Chicagoans still like to project an image of eccentric populism. Only the construction of Navy Pier and a sliver of beachfront at Oak Street has subsequently provided Streeterville some access to the lake.

Then there was Paul Cornell, whose first referendum to build a park south of Hyde Park failed. Meeting his political opponents on their own ground with a proposal to circle the whole city with parks, he won a second referendum. The promise of patronage on a grand scale and the general appeal of parkland for all were an irresistible combination. The resulting green belts at the western edges of the city could not confine subsequent suburban sprawl, but they did provide some breathing room for suburb and city.

Montgomery Ward worked tirelessly and angered many of his

own friends in his successful efforts to improve Grant Park. Yet he was so confirmed in his view of a lakeside without structures that only exhaustion ended his opposition to Marshall Field's natural history museum. Ironically, Ward's opposition is conceded to have vastly improved the siting of the museum.

At first sight, all these ironies seem only to affirm Chicago's image as a blue-collar town presided over by parasitical politicians and predatory businessmen. But Wille looks well past this image to recognize the fundamental dilemmas of public space in Chicago. How to civilize, yet democratize? How can one incorporate into public space a higher aesthetic while also promoting mass usage and participation? How can public space provide a model of civility if it must be achieved by partisan appeals? These American aspirations are brought into stark confrontation on Chicago's lakefront.

Yet as Wille shows us they are not impossible aspirations. One must read her closely, however, for the footwork on both sides is little short of Byzantine. There is always some lurking proposal for private use clothed usually in high-sounding rhetoric about either "civic greatness" or "popular" demand. Straightforward opposition may be the least reliable strategy. Sometimes it's better simply to review the police or financial records of the more interested parties. Then again, it may be helpful to resort to the courts or state legislature if they are thought to include some untouchable advocate of the people. That failing, another option is to propose an alternative plan, one far superior, better documented, and more newsworthy. It may have to include also a measure of patronage. Chicago land-grabbers have been known to retreat if left with enough "cush." Even so, such plans have a way of expiring after the next mayoral election. The problem is to nail them down before the fatal election.

On their own part, those who would make the lakefront into a higher standard of beauty and civility cannot simply oppose any change in land use, for they have their own pet projects and plans. Sometimes, like Montgomery Ward, they appear so elitist and so limited in their appeal as to offend the common man. Thus, they must also package their projects in some compromise that achieves mass support: something for the purist, something to salve ethnic sensitivities, something for the "payrollers."

The defenders of the lakefront work with the dual disadvantage that they have little control over the daily management of the parks and no direct access to funds to accomplish their own projects. Their main resources are a great store of public trust and publicity. The use of either involves a treacherous balance. They cannot over promise, they cannot push the panic button every time something goes wrong, and they cannot engage in too much rhetorical invention. Yes, they must be somewhat manipulative, but keep to the higher moral ground. A broad consensus among their ranks must be thrashed out quietly lest they only betray their weakness.

But try as I have to render Wille's story into reduced scenarios, I cannot succeed even approximately. As she tells it, the story is an evolving one in which each side learns from the other in an endless game of moves and countermoves. You must read her and remain vigilant.

I can only touch upon a few of the newer moves in this game since the completion of her book in 1972. Perhaps the grossest abuse of the lakefront has been the extension of the giant exposition hall (McCormick Place) to an annex across Lake Shore Drive on railroad land to the west. The new annex, on a scale similar to the hall to the east, lacks any similar claim to excellence in design, hovers right on the Drive, and has been combined with a vast parking lot for semitrailers to the south. Access to the lakefront has been made more difficult—it was never easy—for a full mile.

The annex went into the ground with surprisingly little opposition. It was not on parkland. Its construction costs, about double those projected, were well spread around, and everybody seemed to be convinced that expansion was inevitable. "At least it's on railroad land where the first one should have been," some said in admitting defeat. Already there is another planned expansion of the exposition hall, still further west. It may be combined with a stadium for the Bears football team, although a new Mayor Daley seems to be cautious. It may not be the big project to begin an election with. He already has a third airport proposed for a site just east of the Indiana state line, the world's highest building just north of the Sears tower, and a ballpark for the White Sox. Ordinarily, that is enough for a Chicago mayoral campaign.

In some ways the most revealing event of the intervening years was the failure to host a 1992 world fair. Like the 1932 world fair, this one would have enlarged the lakefront parks, rectifying the shortage of parkland available between Twenty-second Street and Forty-seventh Street as compared to the areas north and south. Plans were certainly made and a vast extension of parkland was mapped out on paper. Initially there was widespread enthusiasm for the 1992 fair, countered only by some feeble opposition in Hyde Park, well below Forty-seventh Street. Yet as the plans progressed, so many private projects were linked to the fair that it seemed like an afterthought or pretext for commercial development. No one could articulate an idea for the fair beyond a kind of temporary Disneyland. One day it quietly died, not because of concerted opposition, rather because of the overly transparent manipulation in favor of "end uses."

One gets the feeling that a regional recession beginning in the early 1970s, the attrition of an older group of business and civic leaders, and the uncertainties of local political leadership have resulted in a kind of stalemate. It is fairly easy to stop lakefront encroachments, especially if they involve state or federal funding. Partnerships are always vulnerable to defection. But it is extraordinarily difficult to take the initiative to expand the parks or enlarge their noncommercial offerings. Corporate buy outs, firm relocations, and business failures have left the city with fewer movers and shakers who are accustomed to working with one another. Nonprofit civic leaders, in turn, seem to be at a loss for allies and somewhat intimidated by the fear of being "elitist."

Yet things have not quite reached the depths of Wille's "Age of Cement and Convenience." Despite the volatility of mayoral leadership in Chicago, both Mayors Bilandic and Byrne expanded the usage of the lakeside parks. Under Mayor Richard J. Daley, the main aim of park management seems to have been the avoidance of another riot like the ones that occurred in 1919 and 1968. The parks were closed well before sundown and only those indifferent to police control used them freely. Bilandic and Byrne introduced a wide range of activities, marathons, festivals, and special events that repopulated much of the parkland from dawn to dark. These events, however, have grown to the point that during much of the summer portions of the lakefront resemble a suburban franchise row of restaurants. Yet the crowds are genial and diverse. It is possible, of course, that they are a little numbed by high caloric intake, alcohol, and rock-and-roll music. A major art fair in the spring, however, does succeed without much on-site lubrication.

Mayor Washington continued these events and began a planting program giving portions of the lakefront a promise of refuge amidst the softball diamonds, gun club, airport, parking lots, filtration plant, yacht club, and semitrailers. It was much needed. The mechanization of park care and heavy usage have lowered the life expectancy of a tree to about thirty years.

There have been other incremental improvements. The Chicago Historical Museum, the Shedd Aquarium, and the Art Institute have recently expanded their buildings with minimum encroachment on parkland. Of equal importance, the edges along the Chicago River are slowly being recovered for public use and are to be linked up with the lakefront itself. Someday it may be possible to stroll along these river edges for almost two miles between skyscrapers and a waterway that encircle much of the downtown.

But we can never allow these accomplishments and promises to lull us into complacency. Down the road the management of Chicago's lakefront faces both natural and social difficulties. The diversion of water from the lake may have reached a limit. The unknown possibilities of environmental damage leaves us without certainty of what is the higher moral ground. The multiplication of groups staking their separate claims to the lake and its parks worsens these ambiguities. The increasing role of federal and state agencies creates a labyrinthine course of decision making where the consequences are not easily foreseen. Progressively, democratization of the parks is seen as their commodification. They must "pay their way." And they do. Residential development along the better North Side parks is wall to wall, ground to the sky.

Wille's book brings into our collective memory the necessity and the possibility to see our way through these difficulties. She holds the great lakefront park up as a standard to inform future generations of Chicagoans. Chicago's public space has still not fallen to P. T. Barnum's American Standard: The beauty of an object is proportional to the number of people who will pay the fee to see it.

I am pleased that the University of Chicago Press has rescued this book for future generations of my students and other Chicagoans.

Gerald D. Suttles
1990

Foreword

There is a Chicago, one of the great cities of the world, because there is a Lake Michigan. Over the generations, settlers clustered around the lake's southwestern shores and founded a city (and a metropolitan complex) that has become the most dynamic force in the Midwest and a gigantic factor in the life of the entire nation. Lois Wille's brilliant history tells the story of this priceless lake, the work and wealth poured into its reclamation and beautification, and the relentless pressures of growth, urbanization and speculation that threaten to extinguish it.

The vigil to preserve the lakefront has been maintained from generation to generation by exceptional citizens, such as Aaron Montgomery Ward, who placed the public good above selfish comfort. It is this vigil to which the Metropolitan Housing and Planning Council has dedicated itself. There has rarely been a year in the past four decades when the Council and the citizens have not found themselves embattled against marauders, innocent or designing, who would commercialize the lakefront for one purpose or another. In each of these struggles the Council, through such leaders as Ferd Kramer, Joseph Pois, John Baird, Howard Green, Calvin Sawyier, Thomas Nicholson, Robert

Biesel and Rex Bates, has spoken loud and clear: The lakefront is the birthright of Chicago's people. It must forever be kept open, clean, green and free for the public.

As one debacle after another threatened the lakefront, the ultimate insult came when, without warning or public hearings of any kind, the powerful pillars and concrete ramps of the Stevenson Expressway spread down to the shore, obliterating acres of the lakefront park. The Council decided then that, unless the citizens at large knew the story of Lake Michigan, its irreplaceable value and its place in history, and joined the crusade to save their precious heritage, there would, in a relatively short time, be no lakefront at all. Basic legal research for the book was begun by attorney Richard Wexler, then of the Council staff. The legal research was followed by historic and social fact finding by members of the staff and a number of devoted volunteers. The voluminous material turned over to Lois Wille was substantially expanded through her own meticulous research. The Council's profound thanks go also to Norman R. Anderson, who, as a member of the Board of Governors and Chairman of the Communications Committee, worked over a period of several years to see this project through to completion and into the hands of the publisher.

Mrs. Wille, in her vivid story, has produced an extraordinary chronicle of the planning, the vast efforts, the generous endowments, the ecological and economic problems, the politics and pressures that have made the lakefront what it is today. In all of the tugging and hauling over the fate of the lakefront land, it becomes ironically clear that the chief offender has been the people's government itself. The story brings home the message that the fight for the lakefront is never ended. What the lake and our front yard can be and what they will be is up to all of us who live or work in Chicago and its suburbs. Indeed, whether Lake Michigan lives or dies is up to you.

Dale O'Brien, President
Metropolitan Housing and Planning Council of Chicago

Prologue

"Public Ground—A Common to Remain Forever Open, Clear and Free of any Buildings, or Other Obstruction whatever."

No one has done more for Chicago than the three men who wrote those words along the lakefront of their map of the little prairie town on the southern shores of Lake Michigan.

It was 1836, a year before the town incorporated as the City of Chicago. The men—Gurdon Saltonstall Hubbard, William F. Thornton, and William B. Archer—were charged by the state of Illinois with selling unsettled areas to pay for a new shipping canal. But they refused to sell the lakefront.

With those words they made a promise to the people of Chicago. It was a promise that this city, hustler from its infancy, born and nurtured for shipping, trading, and making money, would do what no other city in the world had done. It would give its most priceless land, its infinitely valuable shoreline, to its people. The lakefront would be dedicated to pleasure and beauty, not to commerce and industry. Whenever

Chicagoans gazed on their spectacular shoreline, they would be rich as the barons of the Riviera.

That was the promise. For a hundred years it was kept, more or less, largely because there were people willing to fight for a free and open lakefront. In those years Chicagoans built seven great lakeshore parks, 2,800 acres that are probably the most stunning park system in the world. Other lakefront cities have walls of smoking plants and shipping docks on their shores, mile after mile of warehouses, granaries, and oil storage tanks. Chicago has mile after mile of sand beaches, green lawns, flower beds, and bicycle paths.

In recent decades the promise has not been kept. Large tracts of the people's land have been grabbed away from them and used for parking lots, filtration plants, an airport for owners of private planes, a gigantic convention hall that, with its parking lots and access roads, eats up more than a mile of lakefront parkland.

The worst may be yet to come. An airport in the lake, dear to the heart of Chicago's powerful Mayor Richard Daley, could mean a shoreline crammed with hotels, motels, and expressway interchanges. A wall of highrises could be built along the Illinois Central Railroad's right-of-way, which parallels the lakefront from the Chicago River to 47th Street, virtually barricading the people from the shore. Interstate highways, with their massive double-deck interchanges, could deface the shoreline from the city's northern tip to its southern border.

There is only one way to prevent these disastrous developments: through the kind of citizen action and citizen pressure that created Chicago's magnificent lakeshore parks. Since the day of the city's birth there have alway been people who were willing to fight for an open, clear, and free lakefront. And there have always been people—usually public officials—who had other plans for the shore, plans to obliterate it with buildings and roads. Why? Mainly because to them, it was cheap. It's difficult to find a period in which some bitter struggle over the Chicago lakefront was not heating the courtrooms. But the legal battles were the nicer ones. Men and women have on occasion gone into combat over pieces of this shoreline, even to their death. Men and women have given generously of their genius to enhance it.

There was Daniel H. Burnham, architect and city planner, who wanted to make the lakeshore a chain of lovely lagoons and grassy islands from the city's northern boundary to its southern, all free and open, for all of the people. Chicago made a fine start in realizing his dream; then World War II halted the work. It never started again.

There was Miss Kate Buckingham, who came home to Chicago after a trip to Europe, outraged that people over there didn't take care of their fine old fountains. So she gave a splendid one to Chicago's lakefront, with a large amount of money to keep it in proper shape, for all time.

There were the raucous and greedy men of the turn-of-the-century City Council, who decided they were going to fill the downtown lakefront with an enormous civic center. (Built, naturally, by several of their very good friends.) They forgot that Aaron Montgomery Ward was watching and willing to take on the whole lot of 'em.

There was Mrs. Albion Headburg, strong-willed clubwoman who raised a formidable protest when city fathers were about to tear down the noble ruins of the Columbian Exposition's Palace of Fine Arts. She organized 6,000 other angry women, and the result is the Museum of Science and Industry in Jackson Park.

There was Frederick Law Olmsted, landscape genius who painted with ponds and wooded slopes, green lawns and lake views and curving paths. From sterile sandy ridges and wild grass, he designed the superb complex of Jackson and Washington Parks, linked by the green Midway.

There were the citizens of the new little town of Chicago who met in the First Presbyterian Church on a cold November night and resolved that when Fort Dearborn was sold, the federal government should "take from the center thereof, in a block having four sides . . . one of which shall be fronting upon Lake Michigan" a parcel of land to be "reserved in all time to come for a public square, accessible at all times to the people."

There were the high-powered businessmen, led by a Chicago newspaper, who wanted—and got—an airport for private plane owners, right on the lakefront. Usurping park land. And there were the high-powered businessmen, led by another Chicago newspaper, who wanted—and got—a gigantic convention hall, right on the lakefront. Usurping park land. Ironically, both are built on park land designed by Daniel H. Burnham, who believed that "the lakefront by right belongs to the people."

And "Ma" Streeter, we can't forget you. No one, thank goodness, fought more fiercely for the lakefront. With husband "Cap" at her side, a shotgun in her hands and an ax handy when she ran out of birdshot, Maria Streeter fought bloody battles for the Near North Side sands. She must have known that one day they would hold some of the world's most expensive real estate.

These are some of the men and women who are responsible for Chicago's lake shore: the glories of it, and its tragedies. In general, Chicago has fared well because of them. Few cities in the world have such magnificent stretches of free waterfront, so children could play and gather memories to warm an old age, so men and women could gaze at a horizon and let the waters roll away their cares.

But like the waters themselves, the lakefront is rarely still. There are structures going up, coming down, and occasionally (too rarely) a new green park emerges. There are more plans, wild schemes, lovely dreams. Some men want to make money from the lakefront and some want to keep it free for their children. The years ahead will see bitter battles, like the ones in the past, fought over Chicago's 30 miles of shoreline.

To understand these struggles, and to know how best to react to them, it will help to know the people and the physical forces that made Chicago's lakefront what it is today. And, above all, it will help to recall what Daniel Burnham wrote in 1909 in his great "Plan of Chicago":

The lakefront by right belongs to the people. It affords their one great unobstructed view, stretching away to the horizon, where water and clouds seem to meet. No mountains or high hills enable us to look over broad expanses of the earth's surface; and perforce we must come even to the margin of the lake for such a survey of nature. These views calm thoughts and feelings, and afford escape from the petty things of life. Mere breadth of view, however, is not all. The lake is living water, ever in motion, and ever changing in color and in the form of its waves. Across its surface comes the broad pathway of light made by the rising sun; it mirrors the ever-changing forms of the clouds, and is illumined by the glow of the evening sky. Its colors vary with the shadows that play upon it. In every aspect it is a living thing, delighting man's eye and refreshing his spirit. Not a foot of its shores should be appropriated by individuals to the exclusion of the people. On the contrary, everything possible should be done to enhance its natural beauties, thus fitting it for the part it has to play in the life of the whole city. It should be made so alluring that it will become the fixed habit of the people to seek its restful presence at every opportunity.

Forever Open, Clear and Free

Chapter 1
The Beginnings

On days when the winter wind lashes like a whip at your face, or on April mornings when you want sunshine and heavy wet snow falls from gray skies, it may help to remember that once this was a land of perpetual summer. But that was 400 million years ago, and we would have had to be fish to enjoy it.

The entire Mississippi Valley was covered with warm, shallow seas of the earth's Silurian Period. Coral reefs grew where Chicago now rises, their flanks covered with bright-colored, waving tentacles carrying microscopic food to millions of gaping gullets. Huge creatures called cephalopods, 20 feet long and something resembling octopi, slithered through the warm Chicago sea, dwarfing the 9-foot-long sea scorpions.

The sea left us the limestone bedrock under Chicago, offering a sure foundation to city skyscrapers. In outcrops such as Stony Island, the stone popped through the surface clay. In Thornton and McCook, great coral reefs that turned into limestone have been cracked open and quarried. Other reefs lie just off the lakeshore from 3900 south to 59th Street (5900 south), and from 70th Street to 79th Street. When engineers digging out space for the Kennedy Expressway reached Addison Street, their soft clay turned into stone: the long-buried Logan Square Reef. And, of course, the sea left millions of fossil fragments that have captivated exploring children for generations. On one Chicago Academy of Sciences field trip to the Thornton quarry, children filled their knapsacks and shopping bags with so many pounds of fossils that the bus got stuck in the mud. It needed a shove from a bulldozer to make the trip home.

A million years ago—relatively recent, compared to the age of the Silurian sea—the first of four invasions of northern ice sheets moved into the Chicago region. Each melted and retreated during a warm period. The most recent, the Wisconsin Glacier, arrived about 50,000 B.C.

By 9,000 B.C., our current warm period was melting down the glacier. Ice sheets fell back, raking the earth like a super-bulldozer, dredging ditches and hollows, piling up mounds of rocks, scouring millions of flat acres. The site of Chicago was still 60 feet below the surface of the great glacial lake, but the areas that are now LaGrange and Palos Park and Homewood and Wilmette were emerging from the waters. Farther west, the hills were carpeted with forests of spruce and fir like those of Northern Canada.

Rushing mile-wide glacial rivers tore through valleys, carrying melting waters to the Mississippi River, and the lake level dropped steadily. By 8,000 B.C., mounds of limestone rose from the lake: Worth Island, Blue Island, Stony Island. The Oak Park sandbar was cut by the waves, and then a newer bar emerged that would one day be called Western Avenue.

After another 2,000 years, the lake had dropped to 20 feet above its present level. The plain of Chicago emerged, and the Des Plaines and Chicago Rivers etched their courses.

The lake is still slowly receding, but it's big enough to be the largest fresh-water body entirely within the United States limits: 307 miles long, 118 miles across at its widest point, and more than 900 feet deep.

Today, if you drive toward the lake from the city limits along one of the east-west streets, you will drop down the steps of the old beaches: 20 feet down to the Oak Park sandbar, 20 feet down to the Western Avenue bar, 20 feet down to the Clark Street level.

At the dawn of the Chicago area's recorded history, this was not exactly a spot of striking beauty. Patches of thin, scrubby wood rose gradually above the windswept dune beaches, and then leveled out into a stretch of dry prairie. Three or four miles from the lake, the prairie slipped into a tract of low, swamplike ground 12 miles wide and covered with dank grasses and sticky muck. In the middle of this damp plain lay a shallow depression called Mud Lake. It ran roughly from Harlem and 47th Street to Kedzie and 31st Street.

At its southwestern end, Mud Lake touched the Des Plaines River. At its northeastern end, it touched the south branch of the Chicago River, a clear little stream, its sandy bottom still untouched by garbage and waste. Its banks were lined with bur oak, hickory, cottonwood and maple.

The Des Plaines River ran southwest into the Mississippi, and the Chicago stream ran northeast into Lake Michigan.

And connecting the two, the muck of Mud Lake.

Thus the reason for Chicago.

Chapter 2

"It Would Only Be Necessary to Make a Canal"

They moved slowly south along the shore of the great waters, small bands of Potawatomi hunting families, never staying long enough in one place to see two springs. The Iroquois raiding parties made certain of that.

Eventually they came to the wind-whipped dunes of the southern shores and saw buffalo drinking at the water's edge, deer and wolves racing through the inland woods, turkey and ducks and geese in the marshes. Good hunting, better than any the oldest men could remember. Their cousins the Miami, who had named the water Misch-i-gon-ong, place-of-great-lake, were moving south, away from the shores. And the remnants of their cousins the Illini, weakened by constant wars, were moving west. So the Potawatomi spread across the lake's curving shore and made the lands their own. Their brothers and allies, the Chippewa and the Ottawa—together, they were "the three fires"—were their close neighbors to the north and west.

Visiting friendly tribes and trading with them was not difficult, because many trails left by their predecessors converged at the spot where the lake took the waters of the little river Checagou, named for the powerful wild onion that grew along its banks. (One day the trails would be called Cottage Grove Avenue, Lake Street, Archer Avenue.) By traveling south on the Checagou, and then carrying their canoes on their backs and sloshing several miles through the mud, they could reach the waters of the Des Plaines, and then the Illinois, and then on to the great Mississippi.

In this favorable setting, the old nomadic life settled into a steady cycle. In the summers they lived a family life in cone-shaped, bark-covered lodges along the rivers and streams. (The lake's edge was too windy for comfort, too swampy.) The women grew corn, squash, pumpkins and beans, and gathered the wild rice that grew along the banks of the Checagou. In fall the men moved to temporary hunting and trapping camps, and in spring they went west deep into the prairie for the great communal buffalo hunt.

And so life passed peaceably for the Potawatomi; they grew and prospered in the stability. When the first white men came, they were secure enough to welcome them warmly. Especially the gentle priest who was brought to them by their cousins the Illini.

In the spring of 1673, Jacques Marquette, 36, a French Jesuit, and Louis Joliet, 28, a Canadian-born French explorer, set out in canoes from the French outpost at Mackinac. Father Marquette was looking for new converts, and Joliet wanted more furs and a waterway link between the Great Lakes and the Gulf of Mexico. With their five companions, they followed the lake shore to Green Bay, then took the Fox River to Lake Winnebago, and then the Wisconsin to the Mississippi. When they reached Arkansas, friendly Indians warned them that to go farther would be extremely dangerous. As they headed back north, Indians along the way advised them to try a shorter route, the one they always used: the Illinois River to the Des Plaines, and then a portage across Mud Lake to the little Checagou.

A group of Illini from the Indian village Kaskaskia, near the present town of Utica, went along as guides and introduced the Frenchmen to the Potawatomi living along the Des Plaines and the Checagou—fine hunters who would be excellent suppliers of beaver skins, Joliet learned.

When he returned to Mackinac, his report was enthusiastic: He had found his new fur sources, and also the link between the Great Lakes and the Gulf. "The place at which we entered the lake is a harbor," he wrote, "very convenient for receiving vessels and sheltering them from the wind . . . we could go with facility to Florida in a bark and by very easy navigation; it would only be necessary to make a canal, by cutting through but half a league of prairie, to pass from the foot of the lake to the river which falls into the Mississippi. The bark, when there, would easily sail to the Gulf of Mexico." (Prophetic words. But it would be 175 years before man finally cut that canal and no longer had to haul and tug boats through the reeking marsh of Mud Lake.)

The following fall, Father Marquette, ailing from dysentery, returned to the Chicago portage with two companions to visit his Indian friends. On December 4, 1674, he reached the windy, grass-covered hill at the mouth of the Chicago. He stayed on shore at that point for several days, writing in his journal of the "majestic seas," with colors that varied to give "an air of mirage instead of the vastness of an ocean." In his few days on the grassy hill he killed three buffalo, four deer, three or four turkeys and a partridge—enough food for the winter.

His party moved up the river five miles to Mud Lake and the portage, but his illness and the winter winds forced them to rest until spring. They built a little cabin just west of the present Damen Avenue bridge and Father Marquette said Mass every day. He wasn't lonely, he wrote, because the portage was a busy spot, on the main route of Indian travel. Parties came by to trade furs, eager for his French tobacco. One group of Indians brought a medicine man to treat him, blueberries, corn, pumpkins and a warm blanket-robe made of 12 beaver skins.

In April, Father Marquette and his party reached Kaskaskia.

The Illini received him warmly, but by that time he was desperately ill. He called his flock together ("five hundred chiefs and old men and fifteen hundred youths," he wrote, "besides a crowd of women and children") and preached his last sermon. About 100 Indians escorted him and his two companions back to Lake Michigan. As they neared Ludington, his friends realized he was dying. They carried him ashore and buried him near the mouth of the river that today bears his name. He was 37.

For the next 25 years the Chicago portage was used regularly by French fur traders and missionaries. Bargemen tugging at ropes, pulling small barks laden with pelts, became a familiar sight in the dank grasses of Mud Lake. Potawatomi children liked to stand guard on the little hill at the mouth of the Chicago, hailing the boats as they floated toward them.

But the booming business at Mud Lake collapsed suddenly at the turn of the century. Fox Indians along the Mississippi, angry that the French had given arms to their rivals the Sioux, closed the Chicago portage to white men.

For 70 years little was known of life around the Chicago and the lakeshore. The curtain finally lifted in 1772, with reports of a new trading post on the north bank of the Chicago River mouth, on the grassy hill above the dunes and the water's edge. The Indians tolerated it despite their ban on the white man, probably because the trader who built it was not white. Jean Baptiste Point du Sable, Chicago's first permanent settler, was born in 1745 in Haiti, the son of a free black woman and a Frenchman. He had moved up the Mississippi from New Orleans, building a profitable business in beaver furs, and settled near Peoria with his Potawatomi wife Kittihawa, or Catherine. He opened a post at the mouth of the Chicago because so many Indian trails converged there, and he was convinced the spot had enormous commercial potential. Two years later he moved his wife, baby son and a large group of Indian employees to the Chicago post. His daughter Suzanne was born there in 1775, the first recorded Chicago birth.

Du Sable, six feet tall, husky and dynamic, was a capable and highly impressive man. When the British arrested him as a rebel spy during the Revolutionary War, he was briefly imprisoned at Fort Mackinac but soon released to run a British-owned farm and trading post near Detroit. British Colonel Schuyler DePeyster, commandant at Mackinac, described him in a report as "Baptiste Point de Saible, a handsome Negro, well educated and settled at Eschikagou; but much in the interest of the French."

After his release, du Sable's trading business on the lakefront thrived and grew into a complex of nine buildings, extending from the lake to about State Street, and from the north bank of the river to about Grand Avenue. It included a house elegantly furnished with a French walnut cabinet, featherbeds, mirrors, pictures and candlesticks—extraordinary for a wilderness outpost. It was a super-cabin, the first of Chicago's great lakefront mansions. Du Sable also had a bakehouse, dairy, smokehouse, poultry house, stables and barns, and his assemblage of Indian and French employees was the beginning of continuous community life in Chicago.

When du Sable opened his trading post, the area was under French influence. The English were confined largely to the eastern side of the Alleghenies. With the peace treaty of 1783, the newly constituted United States gained control over lands reaching to the Mississippi. But a series of wars broke out between the Indians, armed with muskets given them by the British, and American troops trying to protect settlers moving west. They ended, at least temporarily, with General "Mad" Anthony Wayne's victory at the Battle of Fallen Timbers in Ohio. Among the peace terms Wayne demanded and got at the 1795 Treaty of Greenville (Ohio) was possession of "one piece of land six miles square at the mouth of the Chikago River, emptying into the south-west end of Lake Michigan." It was the first indication the new American government recognized the strategic location of Chicago.

The treaty and the relative stability it brought to the region meant more business for the prosperous du Sable. But in 1800, for reasons lost to us today, he sold his property to a man named Jean Lalime for about $1,200 and returned to Peoria, where he owned 800 acres. After his wife's death in 1809 he gave his land to her Potawatomi relatives and retired to St. Charles, Missouri, a wealthy and widely respected man. The stone house he built in St. Charles became the state's first governor's mansion. Du Sable died there on August 29, 1818, at the age of 72. He is buried in St. Charles Borromeo Roman Catholic Cemetery.

Eight years after the United States government got the piece of land along Lake Michigan, the War Department built a small military post on the south bank of the river mouth, across from the du Sable establishment. Little Fort Dearborn was not designed to withstand British cannon, but to hold the Indians in check and to break the British-Indian trade that had begun to flourish in the region. Government agents at Fort Dearborn were ordered to undersell the British at any cost, and they did.

Civilian traders began moving in. The most successful was John Kinzie, of Quebec, a domineering man with a quick temper and a sharp tongue, wise in the ways of getting rich in the wilderness. As a young boy he was an apprentice silversmith, but ran away from home and had phenomenal success as a fur trader. He knew Indian languages and customs, and was a friend of a number of powerful chiefs.

Kinzie bought the du Sable compound from Lalime, planted four poplars and two cottonwoods in his yard and put up a picket fence. For decades the trees on the lakefront hill were a landmark.

Two other homes went up west of the Kinzie house along the river bank—the Anton Ouilmette and Thomas Burns cabins. The Charles Lee farm was built on the south branch of the river, near the present Racine Avenue. Like du Sable, John Kinzie prospered selling potatoes, whiskey, tobacco, sugar and salt to visiting traders and trappers. He lent them money and sold them the silver he crafted. Occasionally, he dabbled in the slave trade.

The little community didn't grow fast, partly because of the harsh weather and marshy soil, so difficult to drain, and partly because of the constant fear of attack by Indians armed with British weapons. The Potawatomi became more and more pro-British and hostile to the settlers as the War of 1812 loomed closer. Several of the old chiefs were friendly, but the young braves vowed to exterminate the white settlement.

From early spring, 1812 was a bloody year for the lakefront. John Kinzie and Jean Lalime had one of their frequent shouting matches outside the fort. Lalime wounded Kinzie in the neck, and Kinzie stabbed Lalime to death. Investigating soldiers quickly exonerated the town's leading citizen.

A few weeks later Winnebago Indians, in full war paint and regalia, murdered two workers on the Charles Lee farm. Terrified, the settlers fled to the fort.

When Mackinac fell to the British, the American commander in the west decided Fort Dearborn couldn't be held. He ordered Captain Nathan Heald to evacuate it—after distributing all goods to "friendly Indians" and dumping ammunition and liquor into the river. It was a fatal error. The distribution of goods drew bands of Potawatomi, Chippewa and Ottawa to the fort. The destruction of the ammunition and the liquor enraged them. Famed Indian scout William Wells and 32 Miami arrived to escort the soldiers and settlers to Fort Wayne. For days, Wells and Kinzie and a number of the soldiers argued with Heald, insisting that it would be suicidal to leave the protective stockade with angry Indians all over the countryside. But Heald had his orders. He was to leave on August 15.

On the eve of the departure Black Partridge, a friendly Potawatomi chief, called on Heald and handed the captain a medal that had been given him by the United States government. He said he could no longer wear it because his people were bent upon killing his white brethren.

At 9 A.M. on August 15 the main gate to the fort was thrown open. Out marched the most pitiful procession Michigan Avenue has seen: a column of 100 soldiers and civilians, including 18 children in a covered wagon, many too young to walk. Captain Wells marched at the head, his face blackened—the Indian sign of impending disaster.

The march had reached the present Roosevelt Road when Wells spotted young Potawatomi, Chippewa and Ottawa braves behind the sand dunes. At 18th Street near the beach the Indians attacked. It wasn't much of a contest—there were more than 500 of them. They killed 24 soldiers, 12 civilian men, 2 women and 12 children, scalping and mutilating them and leaving their bleeding bodies on the dunes.

Wells himself killed eight Indians before he was hit. The Indians cut out his heart and ate it on the spot, a sign of respect for a great warrior.

Another 29 soldiers, 7 women and 6 children were taken prisoner. Several of them were killed that night during ritual torture on the lakeshore. Survivors later reported hearing their agonized screams until dawn. Several more died in captivity, but others were ransomed. Captain Heald's wife was bought back for a mule and 10 bottles of whiskey. Later, she provided one of the few records of slavery in Chicago when she petitioned the Court of Claims for compensation for property lost in the massacre. Among the possessions she lost, she said, were "the negro girl Cicely . . . and also her male infant."

Other survivors were the Kinzie family, who had planned to row across the lake to St. Joseph. When the battle began they rushed back to their house. The next day, when Indians fresh from the massacre gathered around the Kinzie home, the half-breed Potawatomi Chief Sauganash—also known as Billy Caldwell—persuaded them to accept gifts of silver and whiskey from Kinzie and leave.

The war ruined John Kinzie. He lost a fortune in goods and was arrested by the British in Detroit. In 1816, when the fort was rebuilt, the Kinzie family returned to their house on the lakefront. His children recalled that the first things they recognized as they neared the site were skeletons from the massacre scattered along the sands.

To help Kinzie re-establish his business, the War Department gave him a job as trading agent for the rebuilt Fort Dearborn. The military owed him a favor. He had paid the Potawatomis $500 in goods to ransom Captain Heald and his aide, Lieutenant Linai T. Helm, in 1812.

John Kinzie had great faith in the commercial possibilities of Chicago, but he did not live long enough to see the settlement prosper. He was buried in 1828 in the fort cemetery at Lake and Wabash, and later moved to the city burial grounds north of North Avenue. When Lincoln Park was developed there, his body was moved to the Kinzie family plot in Graceland Cemetery.

With the peace of 1815, the American government, bankrupt from the war, looked for ways to develop the West quickly. One of the projects on its agenda was that canal across the Chicago portage, first recommended by Joliet in 1673. In 1816 the government acquired a strip of land 20 miles wide, extending 100 miles from the mouth of the Chicago to a point on the Illinois River west of the Fox, from the Potawatomi, Chippewa and Ottawa Indians. The route for a canal was surveyed by Army engineers, but the insolvent Congress passed the buck to Illinois with an act that authorized the year-old state government to dig the canal. The Illinois General Assembly didn't have the estimated $700,000 either and the project was shelved.

The war's end brought stability to the lakefront, and the flow of settlers, traders, missionaries, promoters and speculators increased along the old Indian trails from Michigan and Indiana.

Food never was a problem. In spring, the settlement women made maple sugar. In summer, it was hominy. In autumn, they picked cranberries. There was abundant game and fowl, berries and beans from the gardens along the river. Life was so peaceful that in 1823 the soldiers moved out of Fort Dearborn, and the little settlement was on its own.

Every September the routine of frontier life exploded into days of wild dancing, drinking and bartering. Thousands of Indians gathered for their annual payments from the federal government for lands they had sold, and every trader in the Northwest arrived to sell them baubles and bangles and cottons and whiskey.

Settlement life also was brightened by the arrival from Detroit of two genial French-descended traders, Jean Baptiste Beaubien and his brother Mark. They were never able to master the frontier game of money-grubbing, but their Gallic gaiety gave the village some spirit, and they taught school and brought the first piano to Chicago and threw dances. Jean, married to a Potawatomi chief's daughter, was a salesman for the American Fur Company. Mark opened a tavern—named the Sauganash for his good friend Chief Billy Caldwell—at what is now the southeast corner of Lake and Wacker Drive.

To out-of-towners from the East, the little village was hardly impressive. When William H. Keating was sent by Secretary of War John C. Calhoun to explore the headwaters of the Mississippi in 1823, he reported: "The appearance of the country near Chicago offers but few features upon which the eye of the traveler can dwell with pleasure . . . . It consists of but few huts, inhabited by a miserable race of men, scarcely equal to the Indians from whom they are descended. Their log or bark-houses are low, filthy and disgusting, displaying not the least trace of comfort."

Lake navigation was dangerous, he reported, and he predicted that sand banks formed on the eastern and southern shores by prevailing winds would prevent "any important works from being undertaken to improve the port of Chicago."

He was right about the troublesome sandbar. It forced the river to curve nearly completely around the Fort Dearborn hill, and enter the lake at the present Madison Street. In the spring of 1818, when the river was dangerously high, the fort commander ordered soldiers to dig a relief channel through the sandbar so it could flow straight out into the lake. But the relief was temporary and the ditch filled during winter gales. Nearly every spring it had to be dug out again. Ships bearing lumber from the North and carrying wheat and Galena lead to the East had to anchor a half mile or more offshore and use small boats to bear their cargo through the shallow lake waters.

Finally, in 1830, the federal government decided to construct a harbor by digging a channel through the sandbar that would be wide, deep and permanent enough to admit sizable ships to the river. But work was delayed because of an old problem: Indian terror. The Sauk uprising along the Mississippi, followed by the return of the Sauk and Fox to northern Illinois, threatened the town. In 1832 young warriors from a number of tribes joined the great Sauk war chief Black Hawk in his effort to stop the advance of the Americans, to make one last brave stand for the ancient homelands. Soldiers returned to Fort Dearborn and settlers fled inside the stockade. More troops arrived from the East, but instead of help they brought another terror: Asiatic cholera. Dozens of men died before the slow-moving ships reached Chicago. The fort became a hospital, and at the fort cemetery at Wabash and Lake fresh graves were dug nightly. But the Indian wars never reached Chicago. Black Hawk and his young rebels didn't get enough support, particularly among the Potawatomi. Troops led by two Indian chiefs who were also noted Chicagoans, Billy Caldwell of the Potawatomi and Shabbona of the Ottawa, stopped the great Sauk warrior.

The troops who survived cholera returned east, full of praise for the fine hunting and farming lands of the Chicago region. They predicted great things would happen there when the harbor and the long-delayed canal were completed.

The harbor was finished in 1833, at a cost of $25,000–for that time, a big federal appropriation. A pier 1,000 feet long was built along the north bank of the river, through the sandbar. Supervising engineer Jefferson Davis–who would be far better known one day as the President of the Confederate States of America–didn't realize what havoc his pier would play with the shoreline south of the river mouth, and the high price the new city of Chicago would have to pay to save that shore.

The canal also moved a step closer. After long negotiations with the state General Assembly and debates over ways to pay for the job, Congress deeded to the state half the land in the 100-mile canal strip, including all that lay within 90 feet of either bank. The state would sell it to pay for the canal.

The year 1833 was significant for another reason. In two years, with the promise of the harbor and the canal, Chicago had grown from about 60 people to more than 150—the quota required by the state for incorporation as a village. The men of the town gathered in Mark Beaubien's Sauganash tavern—where else?—and, with Jean Baptiste Beaubien presiding, they voted to incorporate. Among the trustees they elected was Madore Beaubien, the handsome son of Jean and his Potawatomi wife. He was Princeton-educated and the town's leading beau. Chief Billy Caldwell, the Sauganash, was elected a justice of the peace. This was a truly integrated town, in its infancy. Only one problem, it seemed, had to be settled before the canal lands could be sold and the town could prosper: the threat of the Indians.

That was solved in the manner of the times. The federal government, still barely solvent because of the 1812 war, forced the sale of huge sections of Indian lands for a pittance, moving the Indians to reservations west of the Mississippi, then immediately resold the land to homesteaders for substantially higher prices. Thus, the federal treasury was fed—and the redman was removed to the unsettled wilderness.

In September, 1833, the month after the village was incorporated, more than 3,000 Potawatomi, Chippewa and Ottawa gathered at Chicago to barter away their birthright: 5,000,000 acres east of the Mississippi. They camped on all sides of the town, along sandhills on the lakeshore, on the nearby prairies, in the scrubby woodlands. A visiting Londoner, Charles Joseph Latrobe, watched in fascination and wrote:

Various painted Indian figures, dressed in most gaudy attire, with countenances frequently bedaubed with paint of different kind, red, blue and white . . . . Far and wide the grassy prairie teemed with figures; warriors mounted and on foot, squaws, horses. Here a race between three or four Indian ponies, each carrying a double rider, whooping and yelling like fiends. There a solitary horseman with a long spear, turbaned like an Arab, scouring along at full speed; groups of hobbled horses, Indian dogs and children, or a grave conclave of grey chiefs seated on the grass in consultation . . . .

It was decided that the government would convey the Indians to new lands west of the Mississippi, pay for their upkeep for a year, and in addition give them these sums: $320,000, to be paid in an annuity of $16,000 for 20 years; $150,000 for shops and houses; $70,000 for education; $125,000 in goods; and $110,000 to compensate those who had to give up houses, farms and stores in the Chicago area. In addition, white settlers were given $175,000 "to satisfy claims made by them against the tribes." Hundreds, of course, immediately claimed the Indians owed them substantial amounts. The influential got their claims.

The first payment to the Indians was made on the spot, and the traders were there to profit. An angry observer, Henry Van Der Bogart, wrote: "Many who had three or four blankets the day before yesterday were naked. They will give anything they have for whiskey and as soon as they are drunk they are stripped to the skin by the whites. Such infernal villainy would make the devil blush."

A contingent of Indians left for the Mississippi that fall, and more in 1834. In August, 1835, the remainder gathered in Chicago for the redman's final farewell. They were determined to make it memorable. Easterners who had arrived for the canal land sales watched in horror as 800 braves staged a good-by dance, wearing only loin cloths and a profusion of paint in brilliant colors. Their long hair was gathered in scalp locks on the top of their heads and wildly decorated with hawk and eagle feathers. Led by a band of musicians beating on hollow vessels with sticks and clubs, the dancing line moved slowly, stopping at each settler's house to leap and howl and brandish tomahawks. From the windows of the Sauganash, John D. Caton, a future Chief Justice of the Illinois Supreme Court, watched the dancers and wrote that "their eyes were wild . . . the countenances had assumed an expression of all the worst passions which can find a place in the breast of a savage . . . . The dance consisted of leaps and spasmodic steps, the whole body distorted into every imaginable unnatural position . . . . It seemed as if we had a picture of hell itself before us."

Among the departing dancers were some of the town's leading citizens: Chief Billy Caldwell, who had saved the settlement from Indian raids, for whom the Sauganash was named and whose frame house at the present State Street and Chicago Avenue was the show place of the settlement; Alexander Robinson, or Chechepinqua, another half-breed Potawatomi chief who had a store and tavern at Lake Street and the river; and young Madore Beaubien, who two years before had been elected a village trustee.

In them, the Indian blood was more powerful than the white. And the United States government, still wary of Indian blood, was only too happy to pay them for their property and see them on their way.

The procession ended with one final fierce leaping, howling dance in front of Fort Dearborn—soon to be evacuated, now that they were leaving. Then the Indians headed west, and Chicago saw them no more.

In 1962, anthropologist James A. Clifton noticed on a roadmap that there was a Potawatomi reservation just north of Topeka, Kansas. He had studied the woodlands Indians, but knew of no Potawatomi on the Kansas prairie. He found a small settlement of dilapidated farms, and was startled to see an old religious ceremony in progress conducted in the ancient Algonkian language. Gradually, as he came to know the Indians, he made a surprising discovery. "These Potawatomi were the descendants of the very band that had carried out the Fort Dearborn Massacre," he wrote in the spring, 1970, *Magazine of the Chicago Historical Society.* "The prairie band in Kansas has never forgotten Chicago, nor their victory at Fort Dearborn."

And among their cherished possessions, he saw some handcrafted silver—with the mark of the wily old Chicago pioneer John Kinzie.

Some of the Kansas Potawatomi, Clifton wrote, commute to Topeka to work. Others subsist on rent from their farms. And many live on old age benefits, welfare payments or odd jobs. They are a poor people. But the city they left, on the land that once was theirs, prospered more spectacularly, in a shorter time, than any on earth. The boom began in 1835, when the commercial promise of the lake seemed about to be realized. And, as that prospect rose, a few farsighted people began to look at the lakeshore in another light: as a priceless spot for recreation and refreshment.

1. Fort Dearborn, built in 1803 at the mouth of the Chicago River, on the south bank, offered refuge to settlers during Indian uprisings. Detail of painting by C. E. Petford, courtesy of the Chicago Historical Society. ICHi-03039

2. Chicago in 1820, with farmers, fur trappers and Indian traders once again at peace. Detail of lithograph by C. Inger (1867), courtesy of the Chicago Historical Society. ICHi-18310

3. View of Chicago from Lake Michigan in 1830. Detail of a lithograph from A. T. Andreas, "History of Chicago" (1884), courtesy of the Chicago Historical Society. ICHi-05632

Chapter 3

The Land They Labeled "Free"

Boom town! Hotels and boarding houses jammed, covered wagons dotting the prairies like overgrown daisies, flimsy timber houses rising from the mud, no time to build roads. The word had spread all over the East that Chicago had a great new harbor, Chicago was about to have a shipping route connecting the Gulf with Lake Michigan, Chicago had no more Indians, and Chicago would make you rich. Every hustler east of the Mississippi wanted a part of it. Steamboats from Buffalo, puffing through the new Erie Canal, brought hundreds of hopefuls each week. The busiest place in town was the United States Land Office on Lake Street near Clark, primed with several million acres of ex-Indian land to sell.

From a little muddy village of about 200 in 1833, Chicago had grown into a big muddy village of 3,265 in 1835.

Lots along the proposed canal that would link the Chicago River to the Des Plaines and the Illinois had been up for sale since 1830, after Congress deeded them to the state. But there was little demand for them until 1835, when the Illinois legislature finally was able to get financial backing for the canal from eastern bankers.

Speculators who bought canal lots early in the year resold them a few months later at two or three or four times the original price. By early 1836, they were making a 25 per cent profit in a few days. The crooks were around, too, selling lots that didn't exist, or lots nowhere near the Chicago area.

Men with a little capital and a lot of hustle grew rich, men whose names would be associated with Chicago for decades: Ogden, Hubbard, Wentworth, Gurnee.

The elegant easterner William B. Ogden arrived reluctantly from New York in 1835 when he was 30, sent by his relatives to get a foothold in the Chicago real estate market. His initial revulsion at the sight of the muddy village dissipated the day he made $70,000 profit on a tract his family had bought for $30,000. Two years later he was the town's top real estate promoter and was elected the first mayor as the village of Chicago became the city of Chicago. The man he defeated was John Kinzie's son, which shows the impact of the East on the prairie town.

John Wentworth, the hot-tempered, 6-foot 6-inch, 300-pound "Long John," arrived in 1836 from New Hampshire and got a job as a newspaper reporter on the *Chicago Democrat.* Eventually he owned the paper and 2,500 acres southwest of town. He also became a Chicago mayor.

Walter S. Gurnee, who came from Detroit in 1835, built a tannery, amassed a fortune, invested it in real estate and made a bigger fortune. He also became a Chicago mayor. (Money and land, it seemed, were the best steppingstones into City Hall.)

English-born Charles Cleaver arrived in 1833, built a slaughterhouse south of the city at 37th Street near the lake, and invested his profits in real estate around his stockyards. Soon he had a great tract of property and turned it into a suburb called Cleaverville. Naturally, he became mayor.

Gurdon Saltonstall Hubbard from Vermont first saw Chicago in 1818, when he was 16 and an agent for the American Fur

Company. He put his earnings into shipping and packing plants, but learned the land craze was an easier way to get rich. An example of how it worked: Early in 1835 he bought 80 acres between Kinzie Street and Chicago Avenue for $5,000, and three months later he sold half of the land for $80,000. For some unknown reason, he never became mayor. But he did serve in the state General Assembly and everyone said he and his wife Mary Ann gave the best dinner parties in town.

The greatest land-grab of all fell through, probably because the man who tried it was old Jean Baptiste Beaubien, too much of a Frenchman to understand Anglo-American hustling. In the spring of 1835 the soldiers were beginning to leave Fort Dearborn. Beaubien had lived for many years in a little house on the military reservation. He found an old law providing for sale of unused public lands at the rate of $1.25 an acre and costs, and somehow convinced the federal land agent that this entitled him to buy all the fort property. He walked away with the deed to 53 acres south of the river mouth—for $94.61. Even in 1835, the land was worth more than 100 times that amount.

He was jubilant, but the rest of the town was torn between laughing at the stunt lovable old Jean had pulled and raging because, after all, it was their prized lakefront harbor land he had bought. When Washington found out what its Chicago land agent had done, the sale was declared illegal in court and Jean got back his $94.61. He was furious, and went to live in the lighthouse on the south bank of the river between Michigan and Wabash. In 1854, at the urging of John Wentworth, Congress passed an act giving the old lighthouse to Jean Beaubien. "And there was not a citizen of Chicago who knew him who ever questioned its propriety," Wentworth later said.

The real miracle of the land craze was that it stopped, for the most part, on the west side of Michigan Avenue. In other waterfront cities, warehouses and wharves and factories were permitted to gobble up land at the water's edge, a wonderfully convenient site for industry. And then, when it was too late, the citizens realized how lovely it would be to have a lakefront park.

Two incidents saved the lakefront of downtown Chicago. One resulted from citizen action and the other from a few memorable words written on a map by the three men in charge of building the shipping canal.

During those hectic days of the 1830s, the men of Chicago held weekly town meetings in the First Presbyterian Church, a little frame building on the swampy southwest corner of Lake and Clark Streets. Usually the talk was of commerce and the canal, but on November 2, 1835, they discussed the future of the Fort Dearborn land, scheduled to be sold in a few years. The possibility that the historic site would be covered with grain elevators and packinghouses upset them, and they gave overwhelming approval to this measure:

Be it resolved: That a grant of the said Military Reservation shall be applied for, upon the express condition that 20 acres, parcel of the said Reservation, to be taken from the centre thereof, in a block having four sides of equal dimensions as nearly as may be, one of which shall be fronting upon Lake Michigan, shall be reserved in all time to come for a public square, accessible at all times to the people.

The Fort Dearborn land (east of Dearborn Street to the lake and north of Madison Street to the river) finally went on sale in April, 1839. Democratic President Martin Van Buren was campaigning for re-election at the time, and worried that people of the West would blame him if rich eastern speculators grabbed away all of the best land from the federal reservation. So the Van Buren administration decided to give Chicago its "public square." When sales maps were drawn subdividing the Fort Dearborn land, the block on the west side of Michigan Avenue from Randolph Street to Washington Street was labeled "public ground," and became Dearborn Park. The land east of Michigan Avenue from Madison Street north to Randolph Street, a marshy plot along the lake, also was labeled "public ground." The Democrats' generous gift of the park didn't do them much good, however. Van Buren lost in 1840 to Whig William Henry Harrison.

For many years Dearborn Park was the city's favorite spot for political rallies and expositions. Prominent Springfield attorney Abraham Lincoln spoke there early in 1856 in behalf of the new Republican Party and its anti-slavery platform. But outdoor rallies and fairs gradually went out of style, and in 1892 ground was broken in Dearborn Park for the Chicago Public Library.

The other lakefront park, much bigger and ultimately more significant, was the gift of the three men named by the state to supervise construction of the canal. The three, Gurdon Hubbard, William F. Thornton and William B. Archer, plotted the canal lands in 1836. They decided not to sell the ground between Michigan Avenue and the lake from Madison Street to 12th Street. It was a narrow strip; during bad storms, the lake rushed up to the muddy street. Yet if the canal commissioners had chosen to sell it, the lakefront land probably would have brought a top price from shipping and packing concerns.

Instead, the commissioners wrote on the lakefront edge of their map: "Public Ground—A Common to Remain Forever Open, Clear and Free of any Buildings, or other Obstruction Whatever."

Sixty years later those words would be quoted again and again in courtrooms by lawyers working for Aaron Montgomery Ward.

The biggest day in Chicago's speculative era was the Fourth of July, 1836: the day that work on the canal finally began. The town's leading citizens crowded onto a steamer at Wolf Point and proceeded up the river to a spot that is now 29th Street and Ashland Avenue. (Some of the town's most disreputable citizens crowded on board, too. They were tied up and secured in the hold during the trip.) Among the crowds gathered at the Ashland Avenue site were hundreds of brawny young Irish and German immigrants, newly arrived to dig the canal. After hours of speeches and predictions of great wealth, and after a noisy fight with children who had dumped mud into the official wheelbarrow, the dignitaries shoveled the first earth for the construction of the Illinois and Michigan Canal. Louis Joliet's 163-year-old dream finally was coming true.

The canal commission went broke a few times during the next decade, throwing thousands out of work. The original estimate of $700,000 grew to $8,000,000. And it turned out that 20 of the first 28 miles were limestone, not nice soft clay. By the time the canal was finished, it was 1848 and the big men like Ogden were saying that what the city really needed was railroads, not a shipping canal.

But the canal did carry great quantities of wheat, corn, pork and beef to the East and the South. Chicago, a city of 20,000 in 1848, boomed again. Within two years, 10,000 new citizens had settled there.

It was not exactly a pleasant place to live. No one had paid much attention to such matters as draining the swampy soil or paving the mud streets. "The surroundings harmonize with the general character of the city—with a few exceptions, resembling a trash can," wrote Swedish immigrant Gustaf Unonius in 1845. "The entire area . . . might be likened to a vast mud puddle. I saw, again and again, elegantly dressed women standing on street corners waiting for some dray on which they might ride across the street. Horses and wagons sometimes sank down in the clay and had to be pulled out with great labor and difficulty."

The Chicago Hydraulic Company, a private venture, built a pumping station and reservoir at Lake Street and Michigan Avenue—one of the first commercial enterprises on the city lakefront—and delivered water to the city through an intake pipe extending only 150 feet into the lake. Human and animal waste drained into the Chicago River, emptied into Lake Michigan and came back through that intake pipe into the city's water supply. As a result, typhoid and cholera plagued the city. In 1854 about 1,700 persons, more than 5 per cent of the population, died in a cholera epidemic.

Those two parks didn't get much attention, either. A few shrubs were planted, but then ignored. On July 13, 1847, the *Chicago Daily Journal* reported that deposits of "unfortunate horses and innocent cattle" which had "paid the debt of nature" and were "denied the rights of civilized burial" lay along the lakeshore for a considerable distance, "rendering what might otherwise have been a delightful promenade at evening into a rank offense to all whose olfactory nerves performed their duty."

The lakefront area, called Lake Park in those days, couldn't be ignored much longer. The river-mouth pier that had been built in 1833 to give Chicago a harbor was changing the shoreline. Sand had begun to pile up on the pier's north side, providing a natural lake fill. But on the south side the shoreline began to erode. During every big storm Michigan Avenue was flooded. The rich had built fine homes on the west side of the street, facing Lake Park, and insisted that the city protect their property. The city asked for money from the federal government to dredge a better harbor and build a breakwater, but Washington said no.

In 1851 a solution was offered. The new Illinois Central Railroad would be happy to construct the breakwater and save Michigan Avenue. All it wanted in return was that precious shoreline.

Chapter 4

The Railroad-on-the-Lake

It was Christmas, 1851, but there wasn't much joy in the Michigan Avenue townhouse of Mayor Walter S. Gurnee. In four days he would have to make the toughest decision of his political career. The Illinois Central's proposal to give the city a breakwater in return for a strip of the lakefront was coming to a vote in the City Council, and he didn't know which way to go.

If he agreed to let the railroad have that land, the richest people in town, his neighbors, would be at his throat for destroying their property values by putting a railroad across the street from their front yards. But if he didn't do something soon to protect that ravished shoreline along Lake Park, the cream-colored stone mansions of Michigan Avenue, including his own home, were going to float away.

The city didn't have the money for the breakwater. There were so many demands right now on the treasury. A new waterworks and a new sewage system would have to be built soon, or the deadly epidemics of typhoid and cholera would destroy the young city. The throngs of immigrants streaming in from Ireland and Germany and Scandinavia and the eastern states would go to other cities if Chicago couldn't keep its citizens alive. And new streets would have to be built to lift the growing city out of the mud and the manure. The Chicago stench was getting as famous as the Chicago granaries and packinghouses. Eventually, too, something would have to be done about making a real showpiece, a landscaped promenade, out of the marshy rubbish of Lake Park across from Michigan Avenue. Philadelphia, New York and Washington had hired city planners and landscape architects to do things like that. But Chicago had been too busy making money by shipping grain, corn, lumber and meat from one part of the country to another. Who had time to stroll in a park, let alone think about making one? Every once in a while one of the newspapers would scold, like the editorial in the *Chicago Daily Journal* complaining that "Chicago should be called the city of the pestiferous odor."

All of these things worried the mayor. But the immediate problem was that shoreline. The new, heavily financed Illinois Central Railroad, with its big-name backers such as Senator Stephen A. Douglas and Representative "Long" John Wentworth, was building a line up from Cairo and wanted a terminal in Chicago, at a spot where it could easily get cargo to and from ships. Douglas already had sold the railroad a strip of his lakefront land beyond the city limits, south of 22nd Street. (Some even said that was why he had argued so eloquently in Washington, persuading Congress to grant downstate lands for the construction of the railroad.) Paul Cornell, who owned a big chunk of property in the little town of Hyde Park, south of the Douglas land, also had sold a lakefront strip to the Illinois Central. The railroad executives said they really preferred a terminal in the western part of the city, near the new industrial belt along the south branch of the river. But now that they had these lakefront right-of-ways, they would settle for a lakefront terminal.

Chicago needed railroads. Gurnee knew that. William Ogden had built one going west to Galena (the forerunner of the

Chicago and North Western), and it was already loaded with orders for those new harvesting machines, the reapers, made by Cyrus McCormick, the young Virginian whom Ogden had bankrolled. Other lines were moving in from the east. This new one would connect Chicago to the south, bringing up fruits and cotton and carrying down grain and lumber.

But, as his neighbors constantly pointed out, they had paid a high price for their Michigan Avenue property because the canal commissioners had promised the lakefront would be "forever open, clear and free." Did he want to be known as the man who messed up that stirring dedication?

The property owners on the north and west sides were just as adamant. They wanted the railroad to build that breakwater, because if the railroad didn't do it the city eventually would have to do it—and raise their taxes.

The council meeting on December 29 went just as Gurnee had feared. Representative John Wentworth came up from Washington to argue in behalf of the railroad, talking of jobs for the jobless, lower prices on fruits and vegetables, a saving in property taxes. Big, booming Long John was persuasive, as usual. The council passed an ordinance giving the Illinois Central a 300-foot wide strip on the lake from 22nd Street to Randolph Street, with permission to construct a trestle that would hold double tracks. The west border of the strip was to be 400 feet east of the west line of Michigan Avenue, which put it yards beyond the Lake Park shoreline.

Maybe it was the beauty of the avenue at Christmastime, in a city where there was so little beauty, that helped Mayor Gurnee make up his mind. Or maybe it was the snow that covered the rubble in Lake Park, for once permitting it to look pretty. Whatever the reason, the mayor confounded the railroad and infuriated most of the City Council by vetoing the ordinance.

Wentworth raged, insisting the veto was drafted by a rival railroad, the Michigan Southern. The mayor, he said, had given in to his wealthy neighbors of Michigan Avenue. (Later, Wentworth admitted he owned a fortieth of Illinois Central stock. But people hated to accuse him of conflict-of-interest, or of anything else. Sometime earlier a critic, Ebenezer Peck, had shouted at rallies that Wentworth was shortsighted and stupid for opposing construction of a city water works. The distinguished 6-foot 6-inch, 300-pound member of the House of Representatives replied by punching Peck and easily flooring him.)

The people of the north and west sides were also angry at the Gurnee veto. Throughout the winter and spring, they pressured their aldermen to reconsider the matter. In June, the council passed the ordinance again, this time with enough support to override a veto.

So it was done. And neither side would ever be very happy about it. The railroad had to give up its hope of entering the city near the industrial section farther west; it would be stymied in court when it attempted to grow; it would have to spend millions to hide its tracks and get rid of its steam engines because it ran through the city's front yard. And the city, when it finally got around to expanding its parks and filling in the lakefront, would be plagued by those railroad tracks. Should the park go over them, under them? How could it get around them? And then, a century after the ordinance was passed, a new fight would erupt: Who owned the air above the railroad tracks? That air, remarkably, would become the hottest real estate in the world.

But in 1852, most of Chicago seemed satisfied with the compromise. The Illinois Central began work immediately on a breakwater of stone masonry between Randolph Street and 22nd Street. That fall, the company bought 73,000 square feet of old Fort Dearborn land north of Randolph along the river mouth from the federal government for $45,000, and filled in the lake about 1,000 feet east to create a big tract for a passenger terminal and train sheds. (This is now the site of the Prudential Building and the Standard Oil skyscraper.)

Because lake traffic was growing so rapidly, the Illinois Central filled in more land north of Randolph Street and built a large slip. It was outside the line of the 1852 ordinance, but the increase in lake transportation was considered crucial to the city's economy. No one complained.

By the late 1850s the lakefront north of Monroe Street was thoroughly industrialized. South of Monroe, it was merely a mess: wharves, piles of boards, rocks and garbage, a wooden sidewalk, a dirt beach. North of Monroe, the railroad tracks fanned into broad freight yards and entered a huge train shed through half-moon doors. Two immense grain elevators sat on the south bank of the river mouth, stretching along the lakefront, not far from the grassy spot where Jean Baptiste Point du Sable had built his trading post. On the north bank of the river, east of Michigan Avenue and opposite the grain elevators, the McCormick Reaper and Mower Works was turning out 10,000 machines a year, shipping them as far as London and Hamburg. Packinghouses and grain dryers stretched from the factory to the lakeshore. Squeezed between these plants were several stores, frame houses and the four-story Lake House, one of the city's busiest hotels. The masts of ships, coming and going and anchored at slips, waved along the river and the lakeshore, carrying so much timber from the north that Chicago became the world's largest center for lumber distribution.

But the most spectacular development of the decade was the railroad. In 1850 only one line, Ogden's Galena and Chicago Union, entered the city. By 1856 Chicago was the focus of 10 trunk lines with nearly 3,000 miles of track; 58 passenger and 38 freight trains entered and departed daily. In a half dozen years, the young city had become the world's largest railroad center.

Each new railroad that came to the city had a profound impact on the surrounding neighborhoods. Yards and depots and warehouses and loading platforms grew up along the tracks, responding to the needs of the companies rather than orderly plans for city development. Once laid, the tracks and their assorted buildings became part of the landscape. City planners would talk for decades of railroad terminal consolidation, but those narrow strips of iron scattered around the central city confounded them all.

Chicago—and the jobs and the wealth it promised to dealers and shippers in grain, corn, lumber, meat and iron ore—was the marvel of the Western world. The curious poured in just to see it—and many of them stayed. In 1852, the population was about 38,700. By the end of 1853, it had grown to 60,000. Four years later it was nearly 120,000.

In the summer of 1856 the first Illinois Central train (a little wood-burning locomotive and two cars) arrived in Chicago from Hyde Park, the country town six miles south. Unfortunately, not a single passenger showed up in either direction. Maybe they were worried about riding above the water on those wooden trestles. But business picked up quickly, and the I.C. began three round trips daily to Hyde Park.

It must have been an intriguing trip. One passenger who came up from Cairo, *London Times* reporter William H. Russell, wrote in 1861:

As we approached Chicago, the prairie subsided into swampy land, and thick belts of trees fringed the horizon; on our right glimpses of sea could be caught through openings in the wood—the inland sea on which stands the queen of the lakes. Michigan looks broad and blue as the Mediterranean. Large farmhouses stud the country, and houses which must be the retreat of merchants and citizens of means; and when the train, leaving the land altogether, dashes out on a pier and causeway built along the borders of the lake, we see lines of noble houses, a fine boulevard, a forest of masts, huge isolated piles of masonry, the famed grain elevators by which so many have been hoisted to fortune, churches and public edifices, and the apparatus of a great city. And just at nine o'clock the train gives its last steam shout and comes to a standstill in the spacious station of the Central Illinois Company, and in half-an-hour I am in comfortable quarters at the Richmond House.

The wealth in produce, lumber and ore brought by the railroads and the ships gave the city government money to attack its overwhelming problem: sewage and contaminated water. In 1856 construction began on a fully enclosed underground sewer system, the second in the world. (Hamburg, Germany, built one in 1852.) In the downtown area, brick sewers were laid in the center of the streets and covered with earth, raising the streets and sidewalks at least 10 feet. In outlying areas the sewers were placed in trenches below street level. As streets were raised in the central city, the main floors of homes became semi-basements. Second floors became first floors, and new stairs and new entrances were built to get into them. Chicagoans liked this peculiar style of two-flats so much they copied it in new homes.

In the early 1850s a new waterworks was built at Chicago Avenue and Michigan Avenue, near the lakeshore. Engineers hoped it was far enough from the river mouth and its polluted sewage to provide safe, clean water. It was, until a bad storm flushed sewage far into the lake and brought typhoid in its wake.

The water problem was to plague Chicago for decades. In the late 1860s a tunnel was constructed to a new waterworks crib two miles out in the lake off North Avenue, and the 20-year-old Illinois and Michigan Canal was dredged deep enough to reverse the flow of the Chicago River into the Illinois River. Thus, the city's sewage went downstate instead of into Lake Michigan. But during heavy floods the river backed up again and headed toward the lake. Sometimes river sewage gushed out as far as the new intake crib, and typhoid followed. Finally, two radical steps were taken: A much larger and deeper canal, the Sanitary and Ship Canal, was dredged and completed in 1900. And the Chicago sewer system was rebuilt with a series of "intercepter" sewers along the lakefront that prevented a flow into the lake even during the worst storms.

The raising of the streets and the furious pace of city growth during the 1850s were a constant source of amazement to

European immigrants. The Swedish minister Gustaf Unonius, who had scolded about Chicago's slime and filth in 1845, saw a different city in 1857. He was pleased that the muddy streets had been paved with planks and stones, that brick and even marble structures were rising in the central city. But what delighted him above all was the moving of old buildings on sled-like runners to outlying districts. He wrote:

> *I have seen even three-story buildings travel down the street . . . seldom drawn by more than one horse. Often the entire width of the street is blocked by a house that is out for a walk . . . . Moving the house does not necessarily mean that those living in it must move out. I have seen houses on the move while the families living in them continued with their daily tasks, keeping fire in the stove, eating their meals as usual, and at night quietly going to bed to wake up the next morning on some other street. Once a house passed my window while a tavern business housed in it went on as usual.*

Shipping and building weren't the only things booming in Chicago. Not all the immigrants and newcomers could find work, so many got their money through burglaries, street holdups and safe-cracking. So many travelers were being robbed and so many stories of Chicago's crime were being broadcast in the East that businessmen feared for their future. "The city is at the mercy of the criminal classes," shrieked the *Chicago Tribune* in 1857.

That was one of the reasons the citizens elected Long John Wentworth mayor that year. They wanted him to clean house. The terrible-tempered giant was the man to do it, and the place where he started was the wildest in town: the lakefront den-of-vice called the "Sands." For years, ever since the pier was built in 1833, sand had been deposited north of the river mouth. Enough accumulated to hold a nest of cheap lodging-houses, bordellos, saloons and gambling dens, clustered on the Sands that nobody owned. Property owners whose homes originally had fronted on the lake claimed title to the Sands that accumulated in their front yards, but had never won their cases in court.

The Sands was as popular with firebugs as it was with farm boys and sailors. Whenever one of the shanties burned, and they seemed to burn every few weeks, small boys and their fathers swarmed out with the volunteer firemen, hooting at the frowsy old "madames" and their fallen ladies as they fled the flames. The *Chicago Tribune* reported that "a large number of persons, mostly strangers in the city, have been enticed into the dens there and robbed, and there is but little doubt that a number of murders have been committed by the desperate characters who have made these dens their homes. The most beastly sensuality and the darkest crimes have had their homes in the Sands, so famous in Chicago police annals."

William B. Ogden, always quick to spot anything likely to make a million, also had his eye on the Sands. He wanted a Lake Michigan terminal and harbor area for his Galena railroad. Ogden bought some property facing the Sands and tried to evict the tavern owners and madames adjoining his land. They refused to leave. Ogden then enlisted the support of his old friend, Long John, urging him on with terrible tales of nighttime at the Sands. The new mayor ordered the residents of the Sands to get off by April 20, 1857, or their houses would be burned down. On the morning of April 20, many of the Sands' most notable residents were watching the dogfights at Brighton track on the South Side. While they were gone, the mayor, the fire department and the county sheriff and his posse descended on the Sands. With shouts and jeers, they pulled the terrified prostitutes out of their shanties and tore apart the flimsy buildings. Fire broke out—although all of the law officers denied setting it—and completed the job of clearance.

The Sands' refugees never returned. They just moved to other parts of the city and went about their business.

Long John retired from office for a few years after that energetic term. He said he was afraid consecutive terms might lead to "the desire to make friends and the fear to make enemies." But he was elected mayor again in 1861. This time he shook up the city by firing the entire police force so he could start fresh. He began rehiring the better officers late that day, but for 12 hours on March 26, 1861, Chicago had no police.

That was about the last bit of frolic for four years. The next month Chicago, hundreds of its young men and its new railroads went to war. Under terms of the land grants to the railroads, the government was entitled to use them as it saw fit in times of war. The Illinois Central, in particular, formed an important lifeline for Union armies. Thousands of soldiers rode south on the railroad as far as Cairo and went on to the battlefronts. So did large supplies of grain, meat and ammunition.

The Civil War meant a loss of trade to the city, but the real horror of it was brought home to the lakefront. Camp Douglas, originally a federal military training camp, became a prison for Confederate soldiers. It covered about 60 acres of the Stephen Douglas property from 31st Street to 33rd Street, stretching five blocks west from the lake. Behind its tall stockade as many as 13,000 wretched men were jammed, sick and nearly starving. Pneumonia killed six a day in one winter, and in 1864 more than a thousand died of smallpox and other diseases.

Like Yankee soldiers at the notorious Camp Andersonville in Georgia, many Camp Douglas prisoners tried to dig their way out. A number succeeded in crawling through the tunnels under the stockades. Chicago lived in fear of these escaped "rebels," but Chicago also pitied the men of Camp Douglas. Clothes and food for the prisoners were collected in churches, and relief committees brought medicines by the wagonload.

With the war's end came an era of legislative graft—nation-wide—that connoisseurs consider without parallel. The Illinois General Assembly kept up with the worst of them. Huge sums were squandered, and wound up in privileged pockets, in the construction of a new state house, an industrial university and the Illinois Southern Penitentiary. Prisoners were leased to private contractors to work as slaves. Bribes for right-of-ways for streetcars drawn by horses went as high as $25,000 per legislator.

In this setting, the Illinois Central came to Springfield to plead for land to expand along the lake. The General Assembly was remarkably agreeable and passed an act for the transfer of "a portion of the submerged lands and Lake Park grounds lying on and adjacent to the shore of Lake Michigan, on the eastern frontage of the City of Chicago," to the Illinois Central Railroad. The company was to pay $800,000 for land valued at $2,600,000 or more, if any price at all could be set for it. The legislators generously provided that the $800,000 should go into a newly created Chicago park fund, so the city could begin to develop a park system—except on its downtown lakefront, of course.

Thus, the railroad would not only get Lake Park, but the entire harbor of Chicago.

Gov. John M. Palmer vetoed the bill, pointing out that the price was ridiculously low, and that it could mean the city would lose its port forever. Chicago newspapers screamed, but the Legislature passed the bill again, over the governor's veto. The violent opposition refused to die, however, and in 1873 the Legislature repealed the act.

But the railroad insisted the act was a contract, and irrevocable. It continued to build piers off its Randolph Street peninsula, beyond the 1852 line, and filled in land between its tracks and Lake Park and between 12th Street and 14th Street for roundhouses and shops. (Later, this filled land became a boon to the city's park system. The filled area south of 12th Street was the site of the Field Museum. Most of the new land created on the lakefront before 1900 can be credited to the Illinois Central, although a park was not what the railroad had in mind.)

The railroad took its case to court and fought for years. It also tried to negotiate a sale with the City Council, but abandoned that attempt because, according to the *Chicago Journal*, the aldermen asked the I.C. to pay them off at a rate of $25,000 for eight men. The *Chicago Tribune* scoffed and suggested the story was highly unlikely because "aldermen, even in Chicago, are cheaper, much cheaper then eight for $25,000."

Finally, on October 28, 1910, the United States Supreme Court ruled that the Illinois Legislature did indeed have the right to repeal the sale and that the lakefront park and the harbor it enclosed still belonged to Chicago.

But when the city finally did decide to lay out a park on the lake, it didn't consider the messy downtown shoreline. It looked north, where the cattle still roamed.

4. “Bird’s-eye view of Chicago as it was before the great fire.” Detail of engraving by Theodore R. Davis, *Harper’s Weekly*, Oct. 21, 1871; courtesy of the Chicago Historical Society. ICHi-05671

REAT FIRE.—DRAWN BY THEODORE R. DAVIS.—[SEE PAGE 990.]

Drake Block.
Chicago *Times*.
Sheridan's Head-quarters.
P., F. W., and C. Depot; and C., A., and St. L.
29. C., Cin., and Ind. Cent. Depot.
30. Chicago and Northwestern.
31. Gas-works.
32. Union Park.
33. Washington Square.
34. Dr. Colyer's Church.
35. Water-works.
36. Lincoln Park.
37. Lake Shore Drive.
38. Chicago River.
39. North Branch.
40. South Branch.
41. Entrance of Chicago River into Canal.
42. Union Stock-yards.
43. Dexter Park.
44. Lake Park.
45. South Side.
46. North Side.
47. West Side.
48. Briggs House.
49. Grain Warehouse, near where the fire first began.

5. Canoeing on Swan Lake, largest of Lincoln Park's three lagoons, c. 1870–71. Stereograph; by Copelin & Malander, courtesy of the Chicago Historical Society.

6. Charred remnants of the Rush Street bridge after the fire of 1871. Photo by Jex Bardwell, courtesy of the Chicago Historical Society.

7. View of Pine Street, now North Michigan Avenue, looking south from the Water Tower after the fire. Stereograph, courtesy of the Chicago Historical Society. ICHi-02805

8. Maria (Ma) and George (Cap) Streeter, who founded a riotous "nation" on 186 acres of lakefront sand. Courtesy of the Chicago Historical Society. ICHi-12593

9. Cap Streeter's shack in his shantytown nation, "Streeterville," 1906. Sixty-three years later, the 100-story John Hancock Center rose on the site. Courtesy of the Chicago Historical Society. DN 3554

Chapter 5

A City Circled by Parks

For a city of nearly 200,000, Chicago in the early 1860s had pitifully little space for outdoor recreation. Its motto was "Urbs in Horto" (City in a Garden), but there was scant garden greenery to be found.

Only four sizable areas—three neat, green oases and one big rubbish heap—had been officially designated as parks. Dearborn Park, the square along Michigan Avenue between Randolph and Washington Streets, was used mainly for speeches and expositions; a city ordinance forbade playing ball or even resting on the grass. Jefferson Park (now called Skinner) was a square block bounded by Adams, Monroe, Throop and Loomis Avenue, a bit of shrubbery and grass amid the broken, filthy streets and wooden shacks of the West Side. People in the congested neighborhood called it their "lungs." Union Park was the showpiece; a tract of about 20 acres east of Ashland Avenue bounded by Lake Street and Warren Avenue, its pretty sloping lawns were intersected by gravel walks, and it had flower beds fringed with elms and maples, ornamental fences, tiny lakes and rustic bridges. The rich merchants and lumber dealers who built homes around this lone beauty spot made certain it remained in impeccable condition.

The city may have had good reason for not spending money on its fourth park, the dumping ground along the lakefront. The legal battles between the Illinois Central Railroad and the city and state over the area called Lake Park clouded its future—so doing nothing at all seemed quite practical.

But a city full of dirty stables, pigsties adjoining homes and ditches filled with animal refuse needed breathing space. People complained about the slaughterhouse owners at 26th Street and Stewart Avenue who, during the packing season, disposed of 120 wagonloads of blood and offal a day. City ordinances prohibited dumping it in the river, but the slaughterhouses broke the law without penalty. Glue factories and distilleries and hide and tallow works sent their refuse to the river and thus into the lake to breed typhoid and cholera.

The city scavengers usually dumped from one neighborhood into another. In 1867 the citizens of the southern village of Hyde Park screamed that Chicago's garbage men were using their town to deposit "every species of filthy and decaying matter, from loads of night soil to dead animals of all descriptions."

Dreadful epidemics raged from spring through fall and, besides the terrible human toll, they created a vexing problem for city officials: Where to put all the dead? The city burial ground, 80 acres along the North Side lakefront, was dangerously crowded.

After the tragic summer of 1854, when 1,700 died of cholera, North Side residents begged the City Council to find another burial ground. They worried about the health hazards of living near a cemetery jammed with disease victims. But the North Side, in those days, didn't have much clout in City Hall. The business families had settled in spacious mansions on the South Side along Wabash, Michigan, Indiana, Prairie and Calumet Avenues and along Washington on the West Side. But north?

Too hard to reach from the produce centers at South Water and Lake Streets and the granaries and lumber yards along the South Branch of the river; steady river traffic kept the drawbridges open nearly all day long, and the tunnel under the river at LaSalle Street wasn't dug until 1871. Except for the working-class families crowded into their modest pine cottages, the only people attracted to the North Side were the squatters that police periodically cleared from the lakefront sands.

But the North Siders had a courageous champion: Dr. John H. Rauch, an angry and public-spirited physician who said they were right to be afraid of the jammed cemetery. He wrote a paper on the hazards and read it to groups of civic leaders: "The patients who died . . . have been buried in the Chicago City Cemetery at an average depth of from three to four feet, owing to the fact that graves cannot be dug deeper on account of the water . . . ." Nearly 4,000 Civil War prisoners from Southern armies were buried there, he wrote, a third of them smallpox victims. "These were brought from Camp Douglas, a distance of nearly six miles, and carried through the very heart of the city, and deposited in a soil already overcharged with decaying animal matter and, as has been already shown, totally unfit by its character and locality for such a purpose. Why it was permitted is beyond [my] comprehension. It may well be challenged whether such an instance of flagrant violation of the laws of health has been anywhere perpetrated, within so recent a date, in any civilized community . . . . There must be at least from 18,000 to 20,000 bodies undergoing decomposition at this time."

This "mass of putrid matter" polluted the air, Dr. Rauch said, assisted by "the foul river" and the undrained streets.

He was particularly outraged because in 1860 the City Council had abolished the city Health Department—certainly one of the most foolish official acts in Chicago history. Its functions were transferred to the Police Department, which had neither the time nor the inclination to do anything about them except operate horrendous "pest houses," where victims of smallpox and cholera were kept until they died. A "dead house," or morgue, was built adjacent to the cemetery at North Avenue and Dearborn Street, near the present site of the Chicago Historical Society.

Dr. Rauch's angry speeches led to the city's first real lakefront park.

His dire predictions of plague moved the city's powerful at the same time they were being moved by pleas for recreation space. The City Council put the two together and decided the cemetery should be moved, divided between the two new private cemeteries of Graceland and Rosehill. And the empty graves should become a park.

The job of moving 20,000 bodies never was completed. Some families refused to have their departed relatives carted away to the new cemeteries. Other graves were forgotten and ignored. In 1932, when workmen excavated for the Chicago Historical Society's building at Clark Street and North Avenue, several bodies were unearthed. In 1970, more were uncovered when the Society's addition was built. Excavators for high rises near the park inevitably turn up more bones. But one grave remains as part of the park. David Kenniston, the last survivor of the Boston Tea Party, is buried there. At 70, he was a soldier at old Fort Dearborn, and manager of the popular Mooney's Historical Museum when he was 110. He died, according to Historical Society records, in 1852 at age 115. Instead of moving him to Graceland, the City Council decided to let him lie and today a boulder marks his grave at Clark Street and Armitage Avenue. Another old settler, however, was moved; John Kinzie now rests in Graceland Cemetery.

On October 21, 1864, the City Council passed an ordinance stipulating that a 60-acre tract at the north end of the city cemetery should become Lake Park. There already was a Lake Park, the messy area along the downtown lakefront, but it had been neglected for so long that apparently the city fathers had forgotten its name.

Six months later the President of the United States was

murdered. Chicago, grieving for the man beloved by the people of Illinois, renamed its new lakefront recreation area "Lincoln Park." The City Council appropriated $10,000 for it, and the monumental job of turning the lakefront into a playground for all the people was begun.

Trees were trimmed, a gate was built and landscape architect Swain Nelson, who had come to Chicago from Sweden in 1855, submitted the prizewinning design for the park. He incorporated the little hills and ridges into his plan, a wise decision; years later scientists discovered those hillocks were part of the ancient sandbar shaped 5,000 to 8,000 years ago by the retreating glacial lake. Nelson drained some of the swampy lowlands and planted grass; others were deepened into lagoons. The cottonwood trees that grew along the ridges were preserved, and walks were laid along the crests. ("This is basically not a man-made terrain," said Dr. William J. Beecher, Director of the Chicago Academy of Sciences, more than 100 years later. "This park still represents the shore of Lake Michigan as nearly as possible as it was before man modified it.")

In the next few years, the park blossomed beautifully. It grew as more of the graves were removed, until by 1871 the original 60 acres had been doubled. Work began on a Lake Shore Drive through the park—a tough job, because the lake relentlessly battered and eroded the shoreline and to the west roamed cows belonging to North Side residents. Road contractors repeatedly impounded trespassing cows, and every evening irate householders came to claim their wandering animals. In 1868 New York's 10-year-old Central Park, designed by the great landscape architects Calvert Vaux and Frederick Law Olmsted, sent a gift to the young park in Chicago: A pair of swans, destined to become the first creatures in the Lincoln Park Zoo. Several years later the Lincoln Park commissioners bought a bear cub for $10, the zoo's first furry beast.

Chicagoans, particularly the North Siders, swarmed to their park as if their lungs were collapsing for lack of fresh air and their eyes burning for a clean sweep of horizon.

Newspapers reported that as many as 20,000 to 30,000 flocked there on sunny Sundays, strolling along the walks on the ancient sandbar, driving "Victorias and broughams . . . over the gravel driveways." The largest of the three park lagoons, named Swan Lake after the New York swans, was filled with small canoes paddled by gentlemen in high stiff collars, their ladies holding dainty parasols and trailing fingers in the water. Little paddlewheel boats floated past, churning apart the images of cottonwoods and willows in the still waters. The newspapers proudly proclaimed that "excellent order generally prevailed," with only a few arrests every weekend for "intoxication, fast driving, disorderly conduct and stealing trees."

The birth of Lincoln Park was only one of several outstanding civic achievements of the postwar years. A remarkable group of men who had first come together during the Civil War days was responsible for the age of enlightenment: judges, bankers, doctors, merchants, attorneys. In 1861 they formed a Union Defense Committee to raise money to equip troops and give food and clothes to soldiers' families. After the war, they turned to the needs of the homefront.

One of them was Dr. John Rauch, who crusaded to move the cemetery. Another was Dr. Edmund Andrews, who crusaded for stricter control of the slaughterhouses. "Along the South Branch [of the river]," he told the City Council, "there are a great number of packinghouses where many hundreds of thousands of animals are slaughtered every year, the filth and offal of which block up with a semi-fluid mass of putrefaction the sloughs and water-courses which empty into the river; during the summer and autumn the condition and odor of the river becomes abominable beyond expression . . . ."

Their cause won the support of Ezra McCagg, head of the State Insurance Company of Chicago and a distinguished art collector, a trustee of the new University of Chicago and the Chicago Academy of Sciences, and a leader of the Chicago Relief and Aid Society—a forerunner of today's welfare system. McCagg circulated petitions to stop burials in the city cemetery and convert it into a park. He was a leader of the joint City Council-Board of Trade committee that drew up plans to cleanse the river by dredging a deeper canal that would draw from the lake and reserve the river flow, plans to drain the streets, and plans to install "intercepting" sewers to receive the filth and pump it away from the lake. (Their plans were not completely realized until 1900, 35 years after the committee first met.)

Another member was John Bice Turner, head of the managing committee of the Chicago and North Western Railway and known as one of the few honest railroad magnates of his time. A third member, Henry Greenebaum, president of the German Savings Bank, was one of 20 young men who broke away from Orthodox Judaism and organized the reform Sinai Congregation in 1860. Another member was Charles C. P. Holden, an alderman, a founder of Northwestern Mutual Life Insurance Co. and a veteran of the Mexican War and the California gold rush. He was the prime force in the City Council behind the campaign to deepen the canal and build new sewers. In December, 1870, he was elected council president—a role that burdened him with an enormous job the following October.

These public-spirited men—McCagg, Holden, Turner, Greenebaum, Rauch, Andrews—worked hard to clean Chicago's water and streets, to develop an effective scavenger system, to set up hospitals and poorhouses, to build a sound school system. In 1865 the City Council ordered that free instruction should be provided for all children over six. Public schools went up in the North, West and South Sides. In areas where there were no separate schools for blacks, Negro children were permitted to attend white schools. But only until their own schools were built. Even so, this was considered remarkably generous and liberal. In 1867 historian James Parton wrote that Chicago public schools were "among the very best in the United States." The school buildings were "large, handsome and convenient," he said, noting that "colored children attend . . . and no one objects, or sees anything extraordinary in the fact."

In 1866 the resourceful band of McCagg and cohorts gave its powerful support to Paul Cornell, the rich south suburban lawyer who was campaigning for a park system.

True, Cornell had a self-serving motive: He wanted to attract investors and developers to his property south of the city. Paul Cornell was, after all, a supremely ambitious man. He had arrived in Chicago in 1847 at the age of 25 from White Creek, New York, with only $1.50 and a brand new law degree. The next day the $1.50 was stolen in Lake House, the big hotel on the river. But Cornell managed to get a job right away with his old professor, John M. Wilson, one of the city's most successful lawyers. When Wilson became a judge, Cornell took over the lucrative practice and invested his savings in real estate south of the city. He sold a chunk of it to the Illinois Central Railroad in return for an I. C. station in his fledgling town, Hyde Park. His next goal was convincing the city to buy the lakefront dunes, stunted trees and marshes south of his village for a park.

A park in swamp-and-sand miles south of Chicago? Insane, said the city fathers. Who needs it? You will, someday, said Cornell. Get it now while it's cheap and vacant.

His argument was sound. Chicago had spread so rapidly and so extensively that there was little open space left within the city limits. Except for the converted cemetery of Lincoln Park, no provision had been made for any substantial amount of park land.

Cornell had investigated various methods of running a park system, and settled upon the New York plan. That city had established a Board of Commissioners of Central Park independent of the City Council. Cornell wrote a bill for a similar independent park commission in the south sector of the city and its southern suburbs, empowered to levy taxes, issue bonds and buy land. He spent the winter of 1867 in the state capital, lobbying for his bill against tough opposition that claimed it was costly, farfetched—and if Chicago wanted parks, let the City Council run them. Cornell fought hard, worked hard, made some promises to land-buying legislators and won. The Legislature passed his bill with the provision that it must be approved by voters in the city and the south suburbs. But then the voters turned it down—in the official canvass, at least. Park supporters said the election was fradulent and demanded a recount, but they didn't get it.

Cornell and the pro-park group didn't give up. Lincoln Park was too far from the West and South Sides of the city, they said, and could be reached only by the North Side horse railway that ran along Clark Street. None of the city's railroads passed near Lincoln. This time, instead of pitching their arguments to the South Side (too obviously a matter of personal enrichment for Cornell), the park group had a new, brilliant concept: a city circled by parks and boulevards.

To complement Lincoln on the North Side, they proposed several parks near the city's west border and several south of the border, all linked by grand avenues.

It was an exciting idea, one that grew out of talks between Ezra McCagg and Frederick Law Olmsted. The noted farmer-journalist-landscape architect had been commissioned by McCagg several years earlier to design a $5,500 monument for St. James Episcopal Cathedral honoring members killed in the Civil War.

The pro-park group went back to Springfield in 1869, this time armed with statements from Olmsted and other highly respected authorities pointing out that Chicago was woefully deficient in public parks. "This is owing, no doubt, to the rapidity with which Chicago has sprung up," said the statement from the Chicago Academy of Sciences. "But it is singular, that with all her characteristic business energy and forethought, she has so far neglected to secure ample grounds for park purposes; but the time has now arrived when it becomes necessary to act, and act in the manner that will not leave her behind, as compared with other cities, in those arts which embellish and render cities attractive as places of abode; in other words, we want, not alone a place for business, but also one in which we can live."

In February, 1869, the Legislature passed bills setting up three independent park commissions for the city's three divisions: south, west and north. Voters in the west and south divisions approved the measure after a vigorous campaign by the park group, but much opposition developed in the north area. We've already got our park, said the North Siders; if we pass this, it just means we'll have to pay a new tax to help out the South and West Sides. Not exactly a public-spirited sentiment, but so strong that the park supporters were afraid to submit the issue to a referendum. As a result, the Lincoln Park Commission didn't have the same firm legal footing as the South and West Commissions. But the judge of the Cook County Circuit Court appointed a commission anyway, as the bill had provided. From the start, there was factional squabbling. The Legislature had stipulated that the park fund was to be built up through property taxes; immediately, each of the three areas complained that it was being cheated because the property in the other two areas was under-assessed, and people there weren't paying their fair share. In particular, the South and West Sides were angry that part of their tax money went to the North Side, which already had its park—and which was assessed at very low rates because most of its housing was so modest.

The City Council, jealous of the commissioners' power, introduced a bill in Springfield to abolish the park boards and give their functions to the city's Department of Public Works. Fortunately, the bill failed.

The South and West commissions began buying park land soon after the 1869 law was passed and there was more squabbling. Real estate developers and politicians who had lobbied for the bill were making killings by selling property at grossly inflated prices, and property owners and politicians who weren't party to the deals were furious.

But at least Chicago was getting some parks. Parks that 100 years later would still be a marvel, treasures that would require constant protection against the land-grabbers, precious open space to be defended by attorneys in courtrooms, by women who flung themselves against parkland trees as chain saws advanced. Too often, the park protectors would lose. And the fight to save the parks conceived in the late 1860s seems destined to be a part of Chicago life in the future, as it has been for a century past.

The year 1869, when Chicago developed its park system, was a momentous one for the lakefront for another reason. The federal government appropriated money for the improvement of a harbor in Lake Michigan at the mouth of the Calumet River, 12 miles south of downtown Chicago. The river flowed from Lake Calumet and the Calumet Sag depression cut by glacial waters 11,000 years ago. Through a channel cut along this fissure, ships could reach the new harbor via the Mississippi River system. The Calumet area also had extensive land for factories and refineries, and access to railroad facilities.

Two years later the first cargo vessels tied up along the wharves of new Calumet Harbor and a real estate boom was underway. By 1874, smoke poured from the tall stacks of new industry near the river mouth, railroads steamed from the north, south and west, and a bustling business center had developed at what is now Commercial and 92nd Streets.

With the growth of the steel industry and the development of chemical and petroleum refineries along the Illinois-Indiana border, Calumet Harbor would take on the job of bringing in the heavy stuff. Here would come the bulk cargo, the grains and oil that required huge storage towers. That meant Chicago's downtown lakefront would be spared the ugliness of those overpowering structures and the rail and truck terminals that go with them. Every so often there would be a move to bring more shipping to Chicago's downtown harbor and talk of a resurgence of the downtown docks at Navy Pier. But to people who love the lakefront, the development of Lake Calumet as a harbor and the gradual decline of the commercial use of downtown Chicago Harbor has been fortunate indeed.

A few months after the three park commissions were set up, Olmsted and Vaux were hired to design a park system. With landscape artist Swain Nelson, they plotted the 500 green acres of Humboldt and Garfield and Douglas Parks, with their quiet lagoons and rose gardens and lily ponds and music courts. Under the guidance of Olmsted and Vaux, the South Park commissioners laid out two giant parks, sandy Jackson on the lakefront and inland Washington, a total of 1,045 acres connected by a 600-foot Midway Plaisance—Olmsted's name for the broad green runway. Originally, they estimated the land would cost $75,000; but through the efforts of speculators, it soared to $3,500,000 by the time the last lot was purchased.

The unique feature of the plans drawn by Olmsted and Vaux was the belt of boulevards linking the parks—South Park (now Dr. Martin Luther King, Jr., Drive) and Drexel to Jackson Park and Washington Park, Garfield and Marshall to Douglas Park, Douglas and Independence to Garfield Park, Franklin and Sacramento to Humboldt Park, Humboldt and Logan and Diversey to Lincoln Park.

The south and west parks were laid out near routes of commuter railroads, and horses pulled big, awkward coaches along planked streets to Lincoln Park—one of 18 horse-drawn "railroads" in the city. The commissioners wanted every resident to be within a half-hour's carriage drive or train ride from a park. The parks, they said, were designed for all of the people of Chicago, and should be an integral part of their lives.

It was a bold and exciting concept and, despite the bickering over costs and real estate scandals, the city was captivated by its soon-to-be new look. The burst of civic energy in the past five years had greatly improved the quality of life; and now there were grand dreams of the future.

No one could foresee the horrible days ahead, when dreams would turn to ashes. But, looking back, it seems that the night of October 8, 1871, was clearly forecast.

By 1871, Chicago was the rail, livestock, grain and lumber center of the world. But the city was a tinderbox.

Nearly 300,000 people lived within the city limits: four miles west from the lake to Pulaski Road, and six miles north from 39th Street to Fullerton Avenue, with a finger stretching along the lake to Diversey for Lincoln Park.

With the exception of the mansions on the spacious lawns along Wabash, Indiana, Prairie, Calumet and Washington Boulevard, the citizens were crammed into flat-faced, flat-roofed frame structures, often two to a lot. In the past 20 years 130,000 people had poured in from Ireland and Germany and Sweden and Norway, two or three families crowding into little cottages lit by kerosene lamps and heated by wood-burning stoves. Just about every house had an adjoining barn for the family cow, goat or chickens—and for hay and kindling wood. Apartment buildings were rare; visitors from the East and from Europe always were amazed to see that Chicago was a city of one-family homes (or, at least, homes that were meant to hold only one family).

Although progress had been made in paving the streets with cobblestones or macadam, two-thirds were still laid with pine blocks fitted into the roadways as if they were bricks. The sidewalks were built of wood almost exclusively. A few downtown buildings, including the State Street shopping area that Potter Palmer was developing, used brick and stone and iron; two of Palmer's most impressive new edifices were the elegant 225-room hotel at State and Monroe that bore his name, and the opulent marblefaced store at State and Washington he built for Marshall Field, Leiter and Co. But, for the most part, Chicago was a city of wood, befitting the world's lumber capital. Even the church steeples were made of wooden frames covered with tin or copper sheathing.

The summer of 1871 had been abnormally dry. Between July and October there were only five inches of rain, about a quarter of the normal amount. The city was dusty and parched; leaves were gray and powdery, trees drooped in the unusual heat. Throughout the summer barn fires erupted and the small fire department (only 185 men) pleaded in vain for more engines, for more fireplugs, for fireboats in the river. The 17 huge wooden grain elevators on the river banks were like kindling wood, the firemen warned, and they wouldn't have a chance to save them if a big fire broke out.

It was spotted first by Dennis Sullivan. He had stopped to see his friends the Patrick O'Learys in their little shingled cottage on the Southwest Side at 137 DeKoven Street about 8 P.M. on Sunday, October 8. Mrs. Catherine O'Leary wasn't feeling well, and the family had gone to bed early. So Sullivan left. He sat for a while on the wooden sidewalk across the street, enjoying the strong southwest winds and listening to the sounds of a party coming from the Patrick McLaughlin cottage in back of the O'Learys'. Mrs. McLaughlin's brother had recently arrived from Ireland and the clan was celebrating.

Sullivan's Sunday evening reverie ended abruptly at 8:45 P.M. He spotted a sheath of flame shooting through the side of the O'Leary barn.

Sullivan ran down the planked street yelling "Fire! Fire! Fire!" at the top of his voice. But for the O'Leary barn and for the city of Chicago, it was too late.

The fire department, exhausted from a large and costly fire that had wiped out four square blocks the day before, could not get equipment to the Southwest Side fast enough. Flames ate their way east, consuming a thousand shanties, houses, mills, granaries. At Van Buren Street the fire vaulted the South Branch of the river. A saloon-and-brothel shantytown called Conley's Patch exploded into flames. The mercantile buildings along LaSalle Street burned more slowly, but just as relentlessly. By 1 A.M. the Chamber of Commerce had fallen; two hours later the courthouse was down, its great bell pealing. The marble magnificence of the Field and Leiter store lay in ruins. The fire raged north across the main stream of the river and badly damaged the new waterworks at Chicago Avenue; but most of the new castellated Gothic watertower, pride of the city and symbol of concern for health, survived. Flames raced after the hysterical families fleeing north to Lincoln Park. Aurelia R. King was the wife of a wholesale clothing merchant. She and her family lived at Rush Street near Erie. In her words:

The wind was like a tornado. I held fast to my little ones, fearing they would be lifted from my sight. I could only think of Sodom or Pompeii, and truly I thought the day of judgment had come. It seemed as if the whole world were running like ourselves, fire all around us, and where should we go? The cry was north, north! So thitherward we ran . . . . We found many fugitives like ourselves . . . everyone asking every other friend, 'Are you burned out?' 'What did you save?' 'Where are you going?' . . . On, on we ran, not knowing whither we went till we entered Lincoln Park. There among the empty graves of the old cemetery we sat down, and threw down our bundles until we were warned to flee once more. The dry leaves and even the very ground took fire beneath our feet . . . . We got into a wagon and traveled with thousands of our poor fellow mortals on and on, at last crossing a bridge and reaching the West Side where we found a conveyance at noon on Monday which brought us out to Elmhurst.

In 24 hours, the fire had killed at least 300, destroyed 17,450 buildings and property valued at $200,000,000 and left 90,000 homeless—nearly a third of the city. But, because of the wind, not a house was touched south of 12th Street; the mansions of the rich survived—and so did the O'Leary cottage, which happened to be south of the burning barn.

Six weeks later, in an official inquiry conducted by City Council President Charles C. P. Holden, Catherine O'Leary wept and said the rumors and newspaper stories blaming her for the fire were vicious lies.

A broken kerosene lamp was found in the ruins of her barn, but she said she had gone to bed early and had witnesses to prove it. She had not milked her cow that night, and as far as she knew her cow had not kicked over the lamp and ignited the disaster. But, she added, there was the McLaughlin party . . .. She heard they had gone to her barn for some milk for whiskey punch.

Not true, said Catherine McLaughlin. At least, she had not seen anyone go out for milk. Besides, her family drank beer, not whiskey-and-milk punch.

And that is as much as Mr. Holden and his investigators could get from the two Catherines.

Much of the city lay in ashes. Tens of thousands needed food and clothes and a place to sleep. And despite the heroic work of the relief agencies that sprouted overnight, it seemed as if Chicago and her ambitions were gone forever. But Joseph Medill's *Tribune*, three days after the fire, ran an editorial that proved to be accurate:

All is not lost. Though millions of dollars worth of property have been destroyed, Chicago still exists. She was not a mere collection of stones, and bricks, and lumber . . . . The great natural resources are all in existence: the lake, with its navies, the spacious harbor, the vast empire of production extending westward to the Pacific . . . the great arteries of trade and commerce, all remain unimpaired, undiminished, and all ready for immediate resumption . . . .

We have lost money, but we have saved life, health, vigor and industry.

Let the Watchword henceforth be: Chicago Shall Rise Again.

The citizens responded. The following month they elected Joseph Medill mayor.

The era of reconstruction dawned, and instead of shelving their dreams for another day, Chicagoans were determined to turn their molten mass into a truly beautiful city. The excitement of rebuilding Chicago attracted some of the nation's finest architects. One of them was Daniel Burnham, 26 years old when he and John W. Root, 22, began their partnership in 1873 and prepared for the age of the skyscraper. Young Louis Sullivan moved in from Boston, convinced Chicago would be the best place to start his architectural career. The South Park Commissioners told Olmsted and Vaux to go ahead with their plans for the park complex, and on the North Side the Lincoln Park commissioners resumed their battle with the lake, erecting piers and breakwaters and seawalls in their determination to build a Lake Shore Drive running through the park from Oak

Street to Diversey. Some dreamers even talked about linking the 2½-mile drive with the 30-mile road up to Fort Sheridan.

The administrative framework for the park system had been completed at the time of the fire, although plans and records had burned and had to be pieced together from memory. In the next two decades, park commissioners were hampered by litigation over land purchases and a cutback in revenue because of the fire, the commercial crisis of 1873 and the slow sale of park bonds. Yet, by fighting for special assessments and working hard to market their bonds, they succeeded in converting nearly 2,000 acres of meadow and wasteland and marsh into gardens and playgrounds that became the pride of the city. Chicagoans were convinced their parks were as fine as any of the famous old ones in Europe—and European visitors agreed. In the first 20 years after the fire, the three park commissions spent $24,000,000—astounding for those years—on a system of 8 big parks, 29 little ones and 35 miles of broad boulevards.

"I should doubt if any city is better provided with parks and boulevards than Chicago," wrote Englishman William Hardman in 1884. "Everything is not yet so completely fixed up as in the Old World, but both parks and boulevards are fine examples of planning and landscape gardening."

Olmsted, who did most of the design work for the south parks, had two quite different goals for them. He studied the series of marshes and ponds and sand ridges along the shore from 56th Street to 67th Street, relieved only by a few stunted oak trees covered with mold, and called it a "forbidding place." The soil was so sandy that water quickly drained away and large plants died for lack of moisture. He proposed that the marshes be deepened into lagoons, with the fill used to elevate the surrounding land. The lagoons would connect to the lake by a channel and a pier at 59th Street. Jackson Park, he said, should be largely water-oriented, with a yacht harbor, winding walks around the lagoons, small bridges, bathing pavilions and plenty of space for boating. Washington Park, stretching from Cottage Grove Avenue to South Park Way (now named Dr. Martin

Luther King, Jr., Drive) and from 51st Street to 60th Street, was to be an open meadow, as natural as possible.

Even before they were finished, the two south parks were jammed with people. By 1874, weekly concerts by the Hans Balatka Philharmonic Society drew thousands who loved his Beethoven, Mozart and Schumann. Jackson Park, only partly cultivated, drew 40,000 picnickers a year. Old photographs of Jackson Park in the early 1880s show children in ruffled gowns and suits and hats sitting under young trees near a pond, canoes in the background; women in bustles and men in vests and white shirts and high collars and derbies playing lawn tennis.

By the mid-1880s the newly finished Washington Park had become the place for the chic to be seen on a fine Sunday afternoon. (During the week, herds of sheep roamed the meadows to keep the grass short.) Every Sunday after church promenades of fashionable carriages moved down Drexel and South Park Boulevards, bearing the wealthy to the elegant new $50,000 clubhouse of Washington Park Race Track just south of the park. Today, the race track has become a suburbanite.

When the first boulevard to Jackson Park was completed in 1873, it drew all the young daredevils who liked to show off their horses. There was so much wild racing down the roadway, and so many complaints from indignant carriage-riders, that the South Park commissioners decreed that owners of fast horses could speed them only between 5 and 7 P.M. on Wednesday evenings and between 2 and 7 P.M. on Friday afternoons. These periods were called "Fast Driving and Concert Days" because they coincided with the free band concerts. Soon as many as 10,000 people were gathering in Jackson on Fast Driving and Concert Days to watch the impromptu races and listen to Strauss and Beethoven.

Thousands more went wild over baseball and tennis, and their new parks gave them what they wanted. Bicycling was a craze, too. But the "outlandish machines," as the *Chicago Times* called them, were banned for a long time from the parks. The Lincoln Park commissioners declared in 1879 that bicycles were foes "to the peace of mind and safety of body necessary to the pursuit of happiness." The cyclists didn't mind; they were more intent on six-day races than wheeling around Lincoln Park, although they did swarm down the boulevards in such huge numbers that the carriage-riders screamed in outrage.

Lawn tennis was introduced from England in 1874, and received enthusiastically. The parks responded by building dozens of courts—40 in the two south parks alone. But one of the first games in Lincoln Park was broken up by an irate taxpayer. He "asserted his equal right to the use of that particular part of the park by sitting down in the middle of the court," the *Chicago Tribune* reported. The angry man continued his sit-in until he was arrested and carried off to jail.

In the 1880s seven baseball diamonds were laid out in the meadows of Washington Park and three in Jackson Park. And at the foot of Randolph Street, in the rubble-strewn mess of forgotten Lake Park on the downtown shoreline, a professional team called the Chicago White Stockings built a grandstand and diamond. More than 8,000 crowded there for the weekend games, arriving at dawn to get seats or good standing space. When the team went on tour and beat Memphis 157 to 1, it was said that 100,000 greeted them on their return. In 1876 the team's new player-manager, Albert G. Spalding, led the White Stockings to the championship of the new National League. Another White Stocking became a famous preacher: Billy Sunday. Later the team became known as Anson's Colts and eventually—because its players were so young—it was called the Chicago Cubs.

Baseball, tennis, boating, cricket, swimming, riding, ice skating, dancing, band concerts—the people of Chicago had them all in their new parks. In Lincoln, oldest and most highly developed, people could play from sunrise to long past sunset. Arc lamps were installed in 1883, and Chicagoans traveled for miles on summer evenings to marvel at their glow.

In 1873 a bathing house was built off North Avenue, but the sandy beaches of Lincoln Park were still many years away. The

North Avenue "beach" was paved with granite block, designed to buttress the impact of the waves and protect the road. Not very comfortable, but bathers in long black tights and striped shirts crowded to the water anyway.

The Lincoln Park Zoo, which began with the two swans from New York and the $10 baby bear, grew with donations of animals from wealthy Chicagoans and a collection purchased by the commissioners from the Barnum and Bailey Circus. By the late 1880s the zoo was extremely popular, and on sunny Sundays 50,000 would visit the animals and then stroll down the shaded walks or sightsee in hired phaetons (lightweight four-wheeled carriages) and ponycarts. The phaeton ride through Lincoln Park had a South Side rival; a phaeton line ran south along Drexel Boulevard to Washington Park, down the Midway to Jackson Park, and then north along Hyde Park Boulevard and Drexel to Oakwood. A delightful nine-mile round trip for 30 cents.

After the fire the beauty of Lincoln Park attracted the wealthy, who had thought there was no place to live but the South Side. The trend north got a big boost in 1884 when Potter Palmer, the shrewd real estate developer, filled in a frog pond and replaced it with a $250,000 house that was even more impressive than the Marshall Field and George Pullman and Philip Armour mansions on Prairie Avenue. The three-story structure with its turreted towers stood at 1350 Lake Shore Drive until 1950, when it was demolished for a high-rise. For a half century it was the most imposing house in the city and the impetus for a luxury-building boom. By the late 1880s Lake Shore Drive and Lincoln Park had become the city's No. 1 tourist attraction. William Archer, a London journalist touring for the *Pall Mall Gazette*, wrote: "You wonder that the dwellers in this street of palaces should trouble their heads about Naples or Venice, when they have before their very windows the innumerable laughter, the ever-shifting opalescence, of their fascinating inland sea . . . ."

By the time Potter Palmer built on the North Side, Lincoln Park had expanded into a 250-acre lakeside playground at a cost of more than $3,000,000. It had also become something of a sculpture garden. The Swedes placed statues of naturalist Karl von Linne and scientist Emmanuel Swedenborg; the Germans, of Ludwig van Beethoven, Johann Friedrich von Schiller and Johann von Goethe. Imposing statues arose of Rene de la Salle, of General Ulysses S. Grant, of General Phillip H. Sheridan on his heroic ride to Winchester, of Ben Franklin gazing fondly at a shaggy dog, of a Potawatomi brave in feathers on horseback. Most outstanding of all was Augustus Saint-Gaudens' masterful statue of a brooding Abraham Lincoln, placed in the park in 1887 as a gift of lumber baron Eli Bates.

"If I have any spare time, I shall spend it in a second visit to Saint-Gaudens' magnificent and magnificently placed statue of Abraham Lincoln," Archer wrote. "Surely one of the great works of art of the century, and one of the few entirely worthy monuments ever erected to a national hero."

While everyone else was having fun in Lincoln Park, the commissioners were having troubles with Lake Shore Drive. It had to be rebuilt a number of times. For every block the Drive moved north, it seemed another block sank into the lake. The section between Oak Street and North Avenue was laid out in water about 300 feet east of the shoreline. Fill for construction came from dredging for the park's south pond and from the new courthouse being built downtown. But the builders did such a sloppy job of filling the lake that pools of water remained, stagnating and breeding millions of mosquitoes. Disgusted North Side residents finally finished the fill job themselves.

About the same time, another piece of lakeshore was being filled, one that would become one of the world's most valuable bits of real estate. But first there would be brawling and bloodshed and wild times that probably could have happened only in Chicago.

On July 10, 1886, a broken-down steamboat called the

Reutan chugged out of Milwaukee on a trial voyage. Its ultimate goal was Honduras, via the Mississippi and the Gulf of Mexico. Its captain, George Wellington (Cap) Streeter, was, as usual, up to no good. The red-haired ex-Union Army soldier, logger, fur trapper, gambler and circus owner planned to load the old tub with guns and make a fortune selling them to South American outlaws. His strapping spouse Maria, or "Ma," was the *Reutan's* ideal first mate. But the two got only as far as a sandbar about 450 feet from the Chicago shoreline, somewhere between Chicago Avenue and Oak Street. The old steamboat was stuck, appropriately, close to the spot where the shanties and saloons of the Sands had flourished 30 years earlier.

Several stormy days built up more sand around the stranded wreck, and a little island took shape. Cap—in his circus ring-master outfit of boiled shirt, striped breeches and tails—laid down some metal and wood and built a causeway to the shore. More sand collected, and Cap had a great idea. He would help nature fill in the lake and create his own city. Cap invited building contractors to dump on the sands. They hauled in thousands of tons of debris from the Near North Side, where the rich were erecting their mansions. And soon Cap Streeter, squatter, had staked out 186 acres of sands and dump heap—barren and dreary, but his very own Promised Land.

The police labeled him a trespasser and the landowners nearby watched anxiously, but no one tried to oust him.

Cap dismantled the *Reutan* in 1889 and built a new home out of an old scow. It was known, rather generously, as The Castle. "All this here land is mine," Cap would proclaim to the curious who came to peer. "I made it."

One day N. K. Fairbank, a wealthy property owner on Lake Shore Drive, walked across the wastes and said, "You've got to get out of here. This is my land. I've got riparian rights."

The leathery Cap Streeter, in his fifties, barked back: "I've got squatter's rights and the right to eminent domain. Now git."

He squirted tobacco juice at Fairbank's feet and produced an old musket loaded with buckshot. Ma waved her ax and a pistol, which went off and clipped the rapidly retreating Fairbank in the heel of his shoe. It was the first shot of the Thirty Years' Streeterville War.

After that Cap claimed allegiance only to the federal government and titled his lands "the District of Lake Michigan." He declared himself governor. To prove it, he began selling beer on dry Sundays, performing marriages and selling lots to old saloon buddies for as low as $1 each. A grubby shantytown sprang up between Chicago Avenue and Oak Street, to the horror of the millionaires building along nearby Lake Shore Drive. The federal government, acting on Cap's application as a Civil War veteran, granted him riparian rights in the lake east of his new home—and Streeterville boomed.

But the millionaires and the real estate developers wouldn't withdraw so easily. Hired goons began skirmishing with Streeter's ragtag army. One memorable day in 1894 Cook County sheriff's deputies barged into Cap's shantytown. Fierce Ma peppered them with birdshot. When they ventured back at night, she and Cap poured boiling water on them. They retreated.

More police raids. One night the Streeters were tipped off and ambushed the attackers. Ma's ax almost severed the arm of one policeman. Men in blue fell like birds under musket fire; nine wound up in the hospital and one lost an eye. The Streeters were unharmed, and a newspaper headlined: "Capt. Streeter and His Joan of Arc Repel the Blues."

In one raid, a policeman was shot through the heart with a lead slug and died. But Streeter convinced the judge he never used anything but birdshot and the murder case against him was dismissed. Another policeman died after one of Cap's buddies pierced him with a pitchfork, but the man pleaded self-defense and was acquitted. By this time Cap and Ma were folk heroes to thousands of Chicago's poor, and law authorities decided they had to get rid of them once and for all—or have a revolution on their hands. In a pitched battle, 100 hired gunmen drove Cap and Ma into the ramshackle Castle. It looked as though the

Streeters were done for. But that night the invaders got drunk on Cap's whiskey stock. He sneaked out, rounded up his friends and a massacre began. Within half an hour 33 wounded invaders, several of whom later died, lay on the bloody lakeshore sands. Streeter dumped them into a wagon and drove them to the Chicago Avenue police station, demanding they be locked up on charges of assault and battery and trespassing.

Nightly battles followed. Once, in Cap's absence, a strongarm squad put the torch to Streeterville's shanties. A frontier gunfighter named John Kirk showed up, and it was rumored he had been hired to kill Streeter. One day Kirk was found shot dead through the heart. Cap Streeter was convicted of murder and sentenced to life in prison.

The shock killed Ma Streeter.

But within nine months mitigating facts about the shooting led to a pardon for Cap and he was back—nearly 70 and full of vengeance. There were new fights, a new Castle, a new Ma Streeter named Alma. But the last battle was near. Cap shot a police captain and lost an eye when a bullet creased his skull. He was spending more and more time in jail. Finally, on December 10, 1918, the authorities succeeded in throwing the old man out for good. The Gold Coast was taking shape by then, and real estate developers trampled over each other to bid for the land that Cap had built.

The grizzled old brawler retired to a hot dog stand on Navy Pier and lived in a houseboat until his death at 84 on January 21, 1921. The big brass he had defied turned out for his funeral, led by Mayor William Hale Thompson. And today, near the site of Cap's Castle, rises the 100-story, $100,000,000 Hancock Center—on a concrete plaza named Streeterville Place.

When his ship ran aground in 1886, Cap Streeter brought back to Chicago a spirit of frontier lawlessness that the city, somewhat successfully, had tried to shed. The Art Institute was drawing acclaim in a building at Michigan Avenue and Van Buren Street, the present site of the Chicago Club; there was a Chicago Historical Society, a Chicago Academy of Sciences, a

Beethoven Society and the Union League Club; Daniel Burnham and John Root were working on their Montauk Building, one of the nation's first skyscrapers. And there were the beautiful parks, grand boulevards, promenades of carriages and broughams, lawn tennis and cricket, bathing, bicycling and baseball.

But in the working-class neighborhoods, crowding and congestion had grown more intense after the fire. Before 1871, the low-income families lived in flimsy cottages. After the fire, apartment buildings popped up all over the city, the dawn of a new kind of urban living. Many of them were ugly frame barracks. They quickly deteriorated into tenements, and as early as 1872 the city health commissioner urged the City Council and Mayor Joseph Medill to pass housing laws controlling the new menace. But the thousands of immigrants arriving every month needed cheap housing. The city government made a few feeble attempts to clean the slums, but without much effect.

Toward the end of the century, the countries supplying most of the new Chicagoans were Poland and Bohemia and Italy. The Poles, more than any other group, strived to own their own homes. By the mid-1880s more than half of the 30,000 in the Northwest Side Polish section lived in houses they owned. Bohemians crowded into a colony called Pilsen between Halsted Street and Ashland Avenue, south of 16th Street; Italians claimed the area along Grand Avenue, west of the river, where the Germans and Scandinavians had lived before moving farther north. New Jewish arrivals, many from Russia, congregated on the West Side in an area of dilapidated shanties and four-story tenements bounded by Polk Street, Blue Island Avenue, Stewart and 15th Streets; it was called, bluntly, the "poor Jews Quarter." (The German Jews who had arrived in pre-fire days moved to better, roomier housing in outlying areas.) The Irish dispersed in various parts of the city, but hung on to their tight little enclave in Bridgeport, their original settlement.

The city's housing shocked Sidney and Beatrice Webb, the

British leaders of social and economic reform who visited Chicago toward the end of the century. So did nearly everything else about Chicago. The government, they wrote:

. . . seems to reach the lowest depths of municipal inefficiency . . . . At present in the Council the Republicans are 30 to 38 Democrats; and the much more important cross-division of honest men versus corrupt is about the same. As some of the honest Democrats refused to coalesce with the Republicans, honest and dishonest, the Democratic majority, mostly dishonest, packed the committees to its liking . . . .

[The streets were] unspeakably bad. [They should have seen them before the great cleanup movements of the 1860s.] The sidewalks are uneven and dilapidated; even when of stone . . . . The streets are nothing but rotten planks in the slums, with great holes rendering it positively dangerous to walk in the dark. The roadways in the crowded business streets are, at best, of the roughest cobbles, mostly unevenly laid, with great holes . . . . In the slum streets they are usually made of wood, merely round slices of tree-stems . . . . This instantly wears uneven, and the unjoined circles of soft wood lie about loose on the mud. Imagine sidewalks and roadways of this sort, the garbage and litter of some of the most crowded slums in the world, in an atmosphere as moist and as smoky as our back country towns, unswept, unwashed, untended from year's end to year's end.

But the Webbs found something to praise in Chicago: *The only decent municipal enterprise—beyond the Fire Department, which is good in all American cities and therefore may be good even in Chicago—is the great system of parks and connecting boulevards . . . . The parks are excellently laid out and kept, those on the South Side in particular.*

Well, Webbs, all of what you said probably was true. But some people in Washington liked Chicago. Liked it so much that in the spring of 1890 they selected it for a singular honor. And, as in the post-fire years, Chicagoans prepared for a new age.

10. Freight yards clutter the downtown lakefront, 1890. Courtesy of the Chicago Historical Society.

11. Michigan Avenue south from Adams Street in the late 1880s. This messy view infuriated A. Montgomery Ward. Photo by J. W. Taylor, courtesy of the Chicago Historical Society. ICHi-05682

12. Michigan Avenue in 1889, south of Jackson Street. Courtesy of the Chicago Historical Society. ICHi-21987

13. View of Chicago's downtown harbor in August, 1891, from the Kimball Building on the southwest corner of Jackson Street and Wabash Avenue. Courtesy of the Chicago Historical Society. ICHi-03148

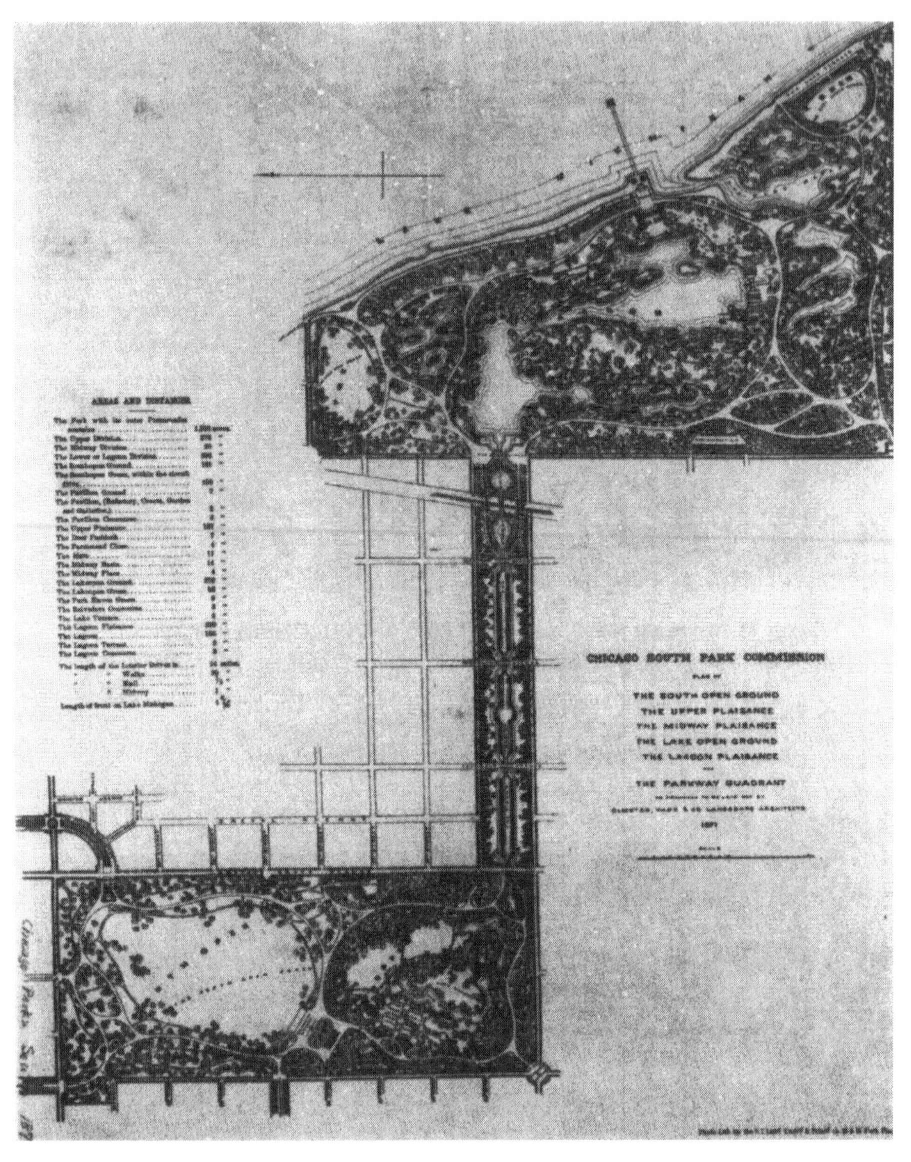

14. The 1871 layout of Jackson and Washington Parks by famed landscape architects Olmsted and Vaux. Engraving, courtesy of the Chicago Historical Society. ICHi-03491

15. View of the Columbian Exposition of 1893 from the administration building, looking east at the Court of Honor. Photo by C. D. Arnold, courtesy of the Chicago Historical Society. ICHi-02530

16. Lincoln Park in 1900. Stereograph, courtesy of the Chicago Historical Society. ICHi-03434

17. Aaron Montgomery Ward, 1844–1913, fought for 20 years to create a lakefront park for Chicago. Engraving from *Historical Encyclopaedia of Illinois,* courtesy of the Chicago Historical Society. ICHi-12788

18. (Below) Grant Park fill, begun in 1901, when dredgings from the river harbor and Chicago Drainage Canal were dumped into the lake. Courtesy of the Chicago Historical Society. CRC-119C

19. (Right) Grant Park fill continues, 1907. Courtesy of the Chicago Historical Society. DN-5234

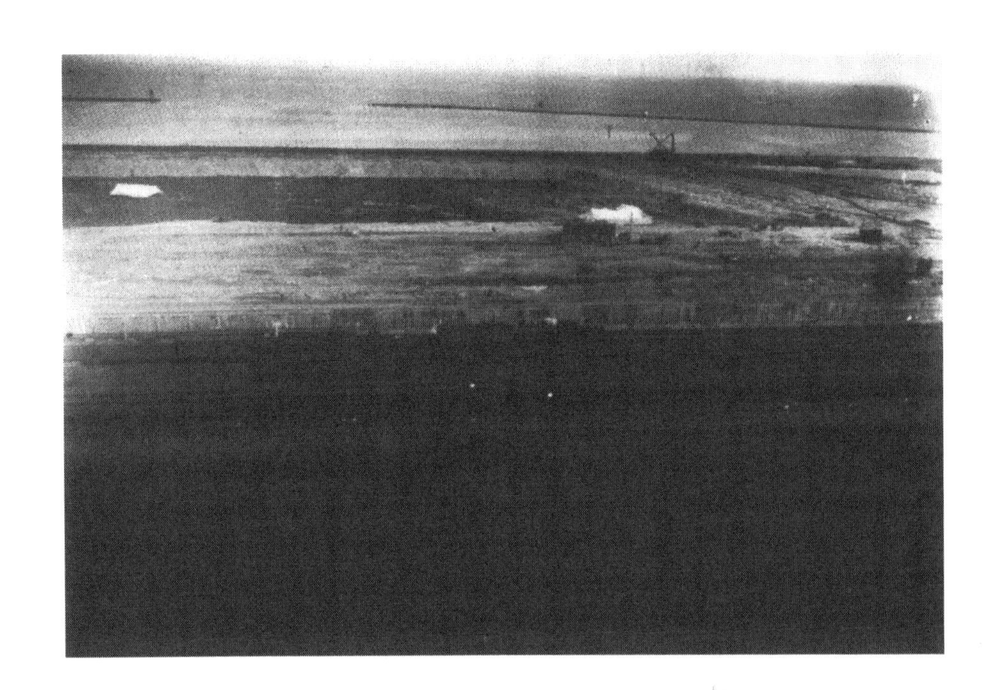

20. Grant Park fill after 1916. Courtesy of the Chicago Historical Society. ICHi-03395

Chapter 6

The Rise and Fall of White City

Chicago had risen from the ashes of a fire that had nearly destroyed her; she had parks acclaimed the finest in the world; her millionaires were generously endowing her museums and her university. She was growing madly, annexing as far as the northern suburb of Rogers Park and the sprawling southern village of Hyde Park, down to 138th Street. Young men were designing tall, graceful buildings that were the wonder of architecture.

And now Chicago thought she could do anything. One disgusted New York newspaper accused the city of "super-voluminous civicism." Charles Dana, editor of the *New York Sun*, popularized a Chicago nickname when he admonished his readers to ignore "the nonsensical claims of that windy city." But that was the spirit that brought the World's Columbian Exposition to Chicago.

Since 1885, Congress had made plans for a gigantic world's fair to mark the 400th anniversary of the landing of Christopher Columbus in the New World. Chicago began campaigning for it immediately, but so did New York, Washington and St. Louis. One way to win, Chicago decided, was to offer more money than any other city. A committee of 300 appointed by Mayor DeWitt Cregier incorporated as the World's Exposition of 1892 and sold $6,500,000 in stock at $10 a share. But the eastern cities lobbied strenuously in Washington. Chicago was a raw new outpost, they said; too new to have traditions and a cultured civilization. To prove it, they gleefully handed around newspaper stories of the adventures of Cap and Ma Streeter.

The city resolved to raise another $5,600,000 for the fair, but was already at the limits of its legal indebtedness. The State Legislature agreed to let Chicagoans vote on a tax increase, and it was easily approved—so eager they were to have the fair in Chicago.

On February 24, 1890, the House of Representatives voted to give the fair to Chicago, and on April 21 the Senate concurred. President Benjamin Harrison signed a bill authorizing the Columbian Exposition for Chicago, and invited the nations of the world to participate.

For nearly 10 months Chicagoans argued over where to put the Exposition. All three areas—North, South and West—wanted it. Some said it might be a good idea to put it on the downtown lakefront, and finally clean up that shabby shoreline.

The fair directors sought the advice of Frederick Law Olmsted. He recommended his Jackson Park—much to the delight of the South Park commissioners, who had spent all their energy and money finishing Washington Park and saw the fair as a means of completing Jackson.

North and West Side residents were disgruntled, but the fair directors and the South Park commissioners argued that the South Side had far better railroad service. Then another argument broke out. The fair directors tried to pressure the park commissioners into unconditionally giving them Washington Park; they had their eye on the lovely meadows as a site for a permanent exposition hall. But the commissioners, men of admirable intelligence and foresight, insisted only Jackson and the Midway could be used—and just for the duration of the fair. Finally, on February 11, 1891, Jackson was formally selected as the site. Opening day was set for May 1, 1893.

Daniel Burnham was named chief of construction and he selected Olmsted as landscape architect. Of the two, Olmsted did a far better job. He laid out broad avenues linking buildings reflected in quiet pools; graceful bridges connected to wooded isles; a pier stretching far out into the lake, arranged so that it served as a landing stage, promenade and breakwater; a fascinating variety of shrubs, flowers and trees.

Burnham commissioned a dozen of the nation's leading architects to create the major buildings, and the noted sculptors Augustus Saint-Gaudens, Daniel Chester French and Lorado Taft to contribute statuary. So much of the country's talent was assembled that Saint-Gaudens exclaimed after one of the planning sessions: "This is the greatest meeting of artists since the fifteenth century." But, instead of promoting the genius of the new generation of architects, Burnham turned back to the Greeks and Romans. Chicago's new buildings, he said, had shown too much "contempt for the classics." This seems strange, since he was one of the outstanding new-school architects. His friends blamed New York architects for swaying Burnham; he had been hypnotized, they said, by the phonies of Fifth Avenue.

The result was a forced neverland—a City Beautiful of ancient facades. Yet it was all handsomely tied together, with a rare harmony and order, and the public thought it was a grand show.

Most of the acclaim went to the Palace of Fine Arts, by New York architect Charles B. Atwood. A reincarnation of ancient Greek art, it rose serenely on the shores of a little lake in the northern part of the park. "The greatest achievement since the Parthenon," proclaimed Saint-Gaudens. And visiting English journalist George W. Steevens wrote that it was "surely as divinely proportioned an edifice as ever filled and satisfied the eye of man."

The Administration Building, by Richard M. Hunt of New

York, had an octagonal gilded dome exceeded in dimension only by the dome of St. Peter's Cathedral in Rome. Below the dome lay a pool with the beautiful Columbian Fountain by Frederick MacMonnies. At one end of the pool a high pedestal rose from the placid waters to support the colossal golden statue of "The Republic" by Daniel Chester French.

Only the terra cotta Transportation Building, by Louis Sullivan and Dankmar Adler, departed from the snowy whiteness of the classic buildings. A rich polychrome decoration on its walls culminated in a great arched doorway intricately ornamented with arabesque and bas reliefs treated with gold and silver leaf. It was the most original of the large buildings, and the only major one designed by Chicagoans. No wonder that Sullivan later said he thought the rest of the fair was "an appalling calamity" architecturally. "The damage it has wrought will last for a half century," he said.

He had wanted the fair to be a showcase for new ideas and techniques, to send contemporary architecture soaring. To see what he had in mind, fair visitors had to go only as far as downtown Chicago, where an impressive array of new-school architecture was developing. Daniel Burnham and his partner John Root had recently completed the 16-story first half of the Monadnock Building, a strikingly unadorned rise of masonry, yet graceful with the curve of its second story and the flow of its windows. Sullivan loved it, calling it "an amazing cliff of brickwork, rising sheer and stark, with a subtlety of line and surface, a direct singleness of purpose, that gives one the thrill of romance."

So did his own Auditorium Building at Michigan and Congress, built in 1889. Constructing a 10-story building so close to the lake created difficult foundation problems. Sullivan and Adler solved them with a massive five-foot thick raft made of layers of timber, concrete and iron beams. Like their Transportation Building, the Auditorium was distinguished by magnificent triple-arched doors and stunning ornamentation. As soon as Sullivan and Adler finished their work for the fair, they began plans for the Chicago Stock Exchange Building, and more of their great arched entrances and tall arched windows.

Even the sternest critics of the fair's architectural backwardness had to love the Midway—probably the greatest collection of fakes in the world. Little Egypt was a sensation dancing on the "Streets of Cairo"; there were Tunisian and Algerian cafes, a Dahomey village, a Persian harem, a German town, an Irish market, a Chinese teahouse, a Hawaiian volcano and Swiss Alps.

"There are gondoliers, real ones from Venice," wrote Gustav Kobbe, a music critic for a New York paper and a magazine writer. "You are made to feel delightfully lazy, lying back and gliding over the pretty lagoons, and imagining yourself in Venice. The illusion continues until your round trip—at an investment of 25 cents—brings you near the little wharf from which you started, and one of your gondoliers remarks, 'Finis! Zee gondolieri lika some beer.' "

And detractors had to admit that the fair passed the acid test: It made money. The 21,000,000 who poured through the turnstiles that memorable summer of 1893 spent enough so that the fair directors could pay a 10 per cent dividend to their stockholders.

But the summer that began with excitement ended with despair. A deep depression gripped the city and the nation. Inside the grounds, all was glitter, gaiety and celebration; outside, jobless men walked the streets and slept in parks. Thousands had moved to Chicago that summer, thinking the fair would somehow lead to steady work. Now they were destitute, straining the meager capacity of the city's private charities. The streets of Chicago were filled with begging men and their children. The fair visitors saw the contrast; they called the fair the White City, and Chicago, the Gray City.

Most of the fair buildings were made largely of wood and plaster, not meant to be permanent. They came down quickly when the fair closed. Only two became lasting parts of the park—the Palace of Fine Arts, which was turned into a museum to hold the Marshall Field Natural History collection, and the

"Convent," or La Rabida. Of all the buildings, this was probably the most appropriate. Donated by the Spanish government, it was a replica of the stone-and-red-tile convent of La Rabida in Halos, Spain, where Christopher Columbus awaited Queen Isabella's decision on his proposed voyage. La Rabida later became a hospital for children with cardiac problems. Even the great Ferris Wheel, the world's first, was taken down. The giant contraption designed by young mechanical engineer G. W. Gale Ferris, of Galesburg, was 250 feet in diameter, with 36 cars that held 60 persons each. Today's ferris wheels are 40 to 45 feet in diameter, and have 12 to 16 two-person cars. The Ferris Wheel was purchased by an amusement park in St. Louis, but eventually sold for scrap metal.

And so the great fair came down. Architecture, and culture in general, had not soared to great new heights. But a new concept had emerged: A planned city, creating relationships between buildings and water and open space that were esthetically pleasing as well as practical and convenient. The layout of the fair had demonstrated the excitement of great vistas, wide walks, greenery and grandeur and elegance in detail. It showed how to plan for pedestrians. It showed how to fit the natural configurations and beauty of the lakefront into a city plan, instead of obliterating them with sprawling, spreading structures. Altogether, the fair in microcosm added up to what the bright new city of the future might be—an idea that Daniel Burnham, in particular, pondered in the coming months.

Chicagoans had spent a big chunk of money fixing their streets and railroads and building an elevated transportation line to the South Side for the fair visitors. They had also grossly overbuilt in surrounding Hyde Park and Woodlawn, where "vacancy" and "for sale" signs stayed for years after the fair closed. But Chicago got a magnificent park in return; at last Jackson was completed in the original Olmsted design. The lagoons became small-craft harbors. The Midway was transformed into a handsome doorway to the University of Chicago. And the beaches, served by excursion boats from downtown, were a relief to thousands on hot summer days. In 1898 the yacht harbor was dredged, and in 1900 the first municipal golf course opened in Jackson Park.

Olmsted couldn't have known, of course, that his lovely park, planned for leisurely walks and canoe rides and carriage rides, would one day be attacked by the automobile. The battle of car vs. lawn and shore, a major threat to Chicago's parks in the second half of the twentieth century, was still decades away.

The *Chicago Tribune's* farewell to the Exposition late in 1893 was both wistful and ironic. The fair was "a little ideal world," the *Tribune* said. "A realization of Utopia, in which every night was beautiful and every day a festival, in which for the time all thoughts of the great world of toil, of injustice, of cruelty, and of oppression outside its gates disappeared, and in which this splendid fantasy of the artist and architect seemed to foreshadow some faraway time when all the earth should be as pure, as beautiful and as joyous as the White City itself."

Chapter 7

"... When You Deal with Montgomery Ward"

One early summer day in 1890 Aaron Montgomery Ward looked out the window of his brand new building on Michigan Avenue and barked out: "Merrick, this is a damned shame! Go and do something about it."

His friend and attorney George P. Merrick got to work right away. And that was the beginning of the end for the shabby downtown shoreline called Lake Park—and the dawn of Grant Park, the city's elegant front yard.

Ward's fight to create a lakefront park took 20 years; it cost him $50,000—more than $200,000 in today's dollars. Toward the end, it nearly broke his strong spirit. There was a sensitive man under that stoical facade. "What do people think I seek?" he asked a friend after a particularly harsh attack on him by newspapers and city officials. "I am only doing all this for the rights of the people. I am not getting anything out of it—except abuse."

He was right. He created what may be the most illustrious park in the world, and was the first to establish a citizen's right to have a voice in what happens to public parks. But in his lifetime he was branded a stubborn, wrong-headed obstructionist—standing in the way of progress. (In those days, like today, there were many who thought "progress" meant "build something.")

Aaron Montgomery Ward was born on February 17, 1844, into a big, down-at-the-heels family in Chatham, New Jersey. The Wards moved to Niles, Michigan, when Montgomery—he never used the name Aaron—was nine years old. A few years later he was apprenticed to a trade. But his father, a cobbler, made so little money that the boy decided to get a better job to help support the family. His sister Nancy Hans later recalled the day he "packed his little satchel" and left home to make his way in the world. He was 14.

"He always was a fighter," she said.

The boy found work in a barrel factory at 25 cents a day and sold corn salve on the side. By the time he was 20 he was manager of a shoe store in St. Joseph and earning the very respectable salary of $100 a month. He had a number of jobs in the next few years: clerk at the Field and Leiter store in Chicago, traveling salesman in the farm towns, junior executive with C. W. & E. Pardridge and Company, the forerunner of Hillman's Foods stores in Chicago. All the while he was planning a new kind of business. He would buy merchandise in large quantities from manufacturers and sell it at low prices—by mail—to the big Midwest rural population that had nowhere to shop but in poorly stocked, overpriced general stores.

Ward started to gather stock in 1871, a bad year for storing anything in Chicago. About the time he was ready to open his business, the fire destroyed the paper collars, calico, winter underwear and other goods he had stashed away in a loft above a stable. He started again, with the backing of two co-workers from the Pardridge Company. In August, 1872, from an office over a barn on Kinzie Street, he opened the nation's first mail order house. "Your money back if not satisfied," the young merchant said on his single-page listing of dry goods. (Some of his bargains: "1 Hoop skirt, 1 Bustle and 1 Hair Braid—$1.00;" "2 Pair Men's Cassimere Pants—$3.00;" "8 Pair Children's Stockings—$1.00.")

Rural families responded. Three years later Ward and his new partner George R. Thorne—the two men were married to sisters—were mailing a 72-page catalog throughout the farmlands of America. Their slogan became world famous: "You Can't Go Wrong When You Deal with Montgomery Ward."

The business prospered so well that by 1890 it had moved to its impressive new home, an eight-story "skyscraper" on the northwest corner of Michigan Avenue and Madison Street, with steam-powered elevators and a marble lobby. Even though it wasn't a retail store, people flocked into the "Customer's Parlor" to relax in the splendor and leaf through the latest catalog.

But upstairs, the boss was raging. The view across Michigan Avenue, toward the lake, turned his stomach: stables, squatters' shacks, mountains of ashes and garbage, the ruins of a monstrous old exposition hall, railroad sheds, a firehouse, the litter of one of the circuses that continually moved in and out, discarded freight cars and wagons and an armory rented out for prize fights, wrestling matches and the masquerade balls thrown by Aldermen Hinky Dink Kenna and Bathhouse John Coughlin for the ladies of the Levee. (They were marvelous parties, said brothel madame Carrie Watson, where "joy reigned unrefined.")

The water between the shoreline and the trestles bearing the Illinois Central Railroad tracks had gradually been filled in. Tons of debris dumped there after the Great Fire of 1871 finished the job. As a result, the area called "Lake Park" had grown so extensive that in 1890 Mayor DeWitt Cregier and the City Council announced plans to clear out some of the rubbish and sheds and build a civic center there—a city hall, a post office, a police station, a power plant and stables for city garbage wagons and horses.

All this on the stretch of lakefront that, in 1836, had been labeled on the Canal Commissioners' plat: "Public Ground—

Forever Open, Clear and Free of any Buildings, or Other Obstruction Whatever."

Ward seemed nearly as offended by the goings-on at the parties as he was at the rubble. The shady ladies dressed like harem girls, champagne was as plentiful as beer, and the leading madames arrived with police escorts. "Chicago ain't no sissy town," said Hinky Dink when some church leaders asked him to tone things down.

Years later, when Ward explained why he began his long fight, he said: "The condition of the park was so unbearable, and the nuisance so vile . . . . The city political gang was allowing it to be used for circuses, dog fights—and even Hinky Dink's fashionable social assemblages."

On October 16, 1890, he filed suit to "clear the lakefront . . . of unsightly wooden shanties, structures, garbage, paving blocks and other refuse piled thereon."

Most city officials and the press assumed Ward was merely angry about the junk in the park, not about the plans for a civic center. Oh well, they thought, he's a fussy fellow; but if he wants it cleaned up, we'll go along with him. They didn't realize that Ward was fighting for open space. He wanted the park cleared—not just of garbage and debris, but of buildings.

The city corporation counsel did see the implications of Ward's suit. He said the city would begin cleaning out the rubbish, but he argued that the open space on the original city maps meant the land was for public use and a civic center would be public. Also, he said, most of the area to be developed had been water on those Canal Commissioners' maps, and therefore not subject to the stipulation that it be "Forever Open, Clear and Free."

A lower court issued a permanent injunction prohibiting the city or anyone else from building on the site, or using it for anything but a public park. Ward, who hated the publicity he was getting and smarted under the charges that he was blocking civic progress, tried to smooth things over with an offer to Mayor Cregier. He would develop a beautiful lakefront park from Randolph Street to 12th Street, at his expense. He wanted it, he said later, "in order that it might be preserved as a park . . . for the benefit of all the people for all time . . . . But my offer was not accepted."

The city appealed the lower court's decision, but in 1897 the Illinois Supreme Court upheld Ward. Down came the temporary post office, the armory, the firehouse.

The City Council was furious, and baffled at Ward's "open space" concept. A downtown lakefront is "no place for a park," complained Alderman William Ballard. "It should be used to bring revenue to the city."

Two buildings were exempted from the order because, as the court had stipulated, all of the adjacent property owners had signed consents permitting construction. One was the Chicago Public Library, built in 1893 at Michigan Avenue and Randolph Street on old Dearborn Park; that site also had been earmarked as public ground on the original city maps. The other was the Art Institute, on the east side of Michigan Avenue at the foot of Adams Street, begun in 1891. One Michigan Avenue property owner, ex-New Yorker Mrs. Sarah Daggett, had refused to sign a consent for the Art Institute. The *Chicago Journal* accused her of representing "a New York clique aimed at crippling" Chicago's preparations for the World Columbian Exposition. Finally Mrs. Daggett's husband signed her name. City officials and the courts, in those dark days before Women's Liberation, decided he had a right to forge her signature and the Art Institute remained.

Years later, Ward said he regretted not fighting the Art Institute construction. He really did want a lakefront completely free of buildings. He would certainly have been upset at recent Art Institute additions. The original consent provided for a building covering only 400 feet along Michigan Avenue; there were some who insisted the additions required new consents from property owners, but the Institute maintained they do not "front" on Michigan Avenue, they "front" in other directions, and therefore they do not exceed the original consent.

After Ward won his case, the city and the Illinois National Guard announced plans to build new armories and parade grounds in the park, but east of the railroad tracks, on land that would be filled in. This site, they said, was exempt from the Ward decision because it was water, not part of Lake Park, when the court made its ruling.

But Ward filed suit again, and again a lower court issued an injunction stopping construction.

This time, Ward had defied the *Chicago Tribune* as well as City Hall. The armories were the *Tribune's* pet project; the depression of 1893 and the labor union struggles that followed—particularly the bloody Pullman strike—had convinced *Tribune* editors that Chicago needed troops close at hand. Long Sunday supplement stories recalled the violent days, and warned of the threat of anarchy. Chicago, the *Tribune* said, was the only world seaport without defense installations as part of its harbor facilities. Ward, quiet, hard-working and rather shy, suddenly was portrayed as a threat to the city's security.

Again, the State Supreme Court upheld Ward. The man who wrote the decision, and Ward's strongest ally on the court, was Justice James Henry Cartwright, son of a frontier Methodist clergyman. He was a tough, forceful man, and played a key role in saving Chicago's downtown lakefront park.

Reluctantly, the City Council conceded that old Lake Park would, after all, be a park. It turned the property over to the enterprising South Park Commission, which quickly set about enlarging it at almost no cost to taxpayers. The commissioners arranged for street sweepings and dredgings from the river harbor and Chicago Drainage Canal to be dumped east of the Illinois Central tracks, creating 50 new acres. Grass and trees were planted, walks constructed, benches installed.

The Illinois Legislature cooperated by passing an act giving the commissioners all submerged lands from Randolph Street to 12th Street as far out as the Chicago Harbor line.

In 1901 the park was renamed after former President Ulysses

S. Grant, a favorite in Illinois because Cairo had been his Civil War military headquarters and he had lived in Galena. (Years later, some said it should have been named after Ward; but at the time it was christened he was considered a public nuisance, if not a public enemy.)

Just when everything was going so well, Ward plunged into his third—and toughest—lakefront battle.

For several years, Marshall Field and the directors of the museum he had created in the Columbian Exposition's Palace of Fine Arts had been looking around for a new home for their collection. They had their eye on Grant Park, not far from where the Art Institute was flourishing.

To make things easier for them—and to circumvent Ward—the state Legislature passed an act permitting cities and park districts to build museums in public parks and levy a tax to maintain the buildings.

Ward, who didn't relish getting into another fight, said he would not try to stop the proposed Field Museum if the park commissioners would agree never to build anything else in Grant Park. They refused. The new law produced a flurry of plans for museums, libraries and monuments in the park. After he had counted 20 of them, Ward filed suit.

Harry G. Foreman, president of the South Park Commission, was so angry he threatened to destroy the new Grant Park. In an interview in the *Chicago Tribune*, he said he was considering stopping all maintenance and "turning the front yard of the city into a rubbish heap" to get even with Ward.

The case was still in the courts when Marshall Field I died in 1906. He left $8,000,000 for the museum; but with the provision that the city must furnish a site, free of cost, within six years after his death.

That provision brought tremendous pressure on Ward. Newspapers, city officials and civic leaders all urged him to pull back, just this once. If we lose that museum, they said, it will be the fault of A. Montgomery Ward. The press called him "stubborn," "a persistent enemy of real parks," "undemocratic." And from his old enemy the *Tribune:* "A human icicle, shunning and shunned in all but the relations of business.

That stung. It was certainly true that Ward was no social lion. He was a multimillionaire, a stocky, energetic man with thick brown hair and a well-tended thick mustache—just the type who could have reigned along with Potter Palmer, N. K. Fairbank, Marshall Field and the rest of the city's elite. But Ward was too much of a private person; he preferred evenings at home with his wife and daughter and the Thorne family to big balls; he liked golf and bred fine horses at his country retreat in Oconomowoc, Wisconsin, but he stayed out of the city's social and cultural affairs. He hated the limelight so much that when he gave to charities and universities and hospitals, which he did generously, he always insisted that there could be no publicity or recognition of any kind. He would not join the boards of settlement houses, but he would give them tons of coal and bushels of food to distribute among the poor, providing they tell no one he had given it.

For this kind of man, the publicity surrounding the Field case was extremely painful. He refused to engage in a public name-calling contest with the directors of the Field estate, although his attorney Merrick wasn't so reticent. "Mr. Field wanted a monument," Merrick said. "And, being a poor man, he couldn't afford to pay for a site. Now it is proposed to secure a site from the city of Chicago by violating a trust, and we don't propose to stand for it."

The battle grew more vicious. Harlow Higinbotham, the president of the Field Museum, remarked to reporters that he really didn't know what motivated Ward, but "I do know he was once a clerk in the Field store . . . ."

Chicago business leaders put pressure on Ward's company. A squad of them took turns following Ward on his business trips to urge him to change his mind. As each returned to Chicago, he would grant an interview to the press expressing his bitterness and disgust with Ward.

The *Tribune* reported that an unidentified man interested in the museum suggested that Ward's customers all over the Midwest write to the firm to complain because "on their visit to Chicago they want to see the museum."

Some of Ward's executives wilted under the pressure, and begged him to withdraw his suit. But not his old friend and partner George Thorne, who offered to help pay the legal bills of the Field fight. Ward said no. He was hurt by the reaction against him, and couldn't understand why no one seemed to appreciate what he was doing, and why.

Again, he offered to withdraw his opposition if all future buildings in Grant Park were outlawed. His offer drew nothing but silence. Then he said he would "subscribe liberally" toward a fund to purchase another site for the Field Museum. The museum trustees didn't want his offer. After a meeting with him, one of them said: "Ward expressed the belief that it was better to have this great tract of land as a place for people to go and lie around on the grass than to make it the pivotal point of Chicago's scheme of beautifying the city—yes, he did, actually."

But the State Supreme Court understood what Ward was fighting for. In 1909 the court again upheld him. In its opinion, written by Justice Cartwright, it conceded that a museum was a proper building to place in a public park. But the issue here was more than a park, the court said. The issue was open space. Ward maintained that the Canal Commissioners' action in 1836, and the city's acceptance of that action, had given him a right to open land along the shoreline—to an unobstructed view of Lake Michigan. The court said he was correct.

The press, city officials and civic leaders reacted with unanimous outrage. For the first time, Ward felt obliged to justify his actions. He granted a long interview to his leading antagonist, the *Chicago Tribune*—the first and only interview of his life. He said:

> *Had I known in 1890 how long it would take me to preserve a park for the people against their will, I doubt if I would have undertaken it. I think there is not another man in Chicago who would have spent the money I have spent in this fight with certainty that even gratitude would be denied as interest.*

I fought for the poor people of Chicago, not the millionaires. In the district bounded by 22nd Street, Chicago Avenue and Halsted live more than 250,000 persons, mostly poor. The city has a magnificent park and boulevard system of some fifty miles, but the poor man's auto is a shank's mare or at best the streetcars. Here is park frontage on the lake, comparing favorably with the Bay of Naples, which city officials would crowd with buildings, transforming the breathing spot for the poor into a showground of the educated rich. I do not think it is right.

Perhaps I may yet see the public appreciate my efforts. But I doubt it.

Time was running out on Marshall Field's six-year provision. The park commissioners insisted there was no appropriate site but Grant Park, so now they sued Ward—an unusual form of condemnation suit asking the court to deny Ward's right to open space in Grant Park. With less than a year to go before Field's deadline, the state Supreme Court for a fourth and final time ruled in favor of Montgomery Ward and open space.

In August, 1911, five months before the deadline, the park commissioners admitted they had lost. They said the museum would be built in Jackson Park, even though that would be too far from the center of population.

Then, with only a month to go, the Illinois Central Railroad stepped in with the ideal solution. The I.C., anxious to shed its image as the leading piece of clutter on the lakefront, offered to give to the park commission the two-block tract of land it had reclaimed from the lake south of 12th Street, adjacent to Grant Park. It had planned to build a new station there, but offered to build it west of the site instead. The offer was accepted.

The site was suggested to the I. C. by Marshall Field's nephew, Stanley Field, who came from England in 1893 at 17 to work in the family store. A week before his death, Marshall Field had told young Stanley to make certain the museum flourished; and Stanley Field devoted the rest of his long life to supervising all facets of the institution. He created a remarkable museum that vividly portrays the history of the world and its creatures. In 1925 he persuaded John G. Shedd, then president of the Field store, to give $3,000,000 for an aquarium near the Field Museum. Three years later Max Adler, a Sears, Roebuck and Company executive, gave $700,000 for a planetarium nearby on a finger of landfill jutting into the lake. The result is one of the finest museum complexes in the world. He was grateful, Stanley Field said years afterward, that Montgomery Ward had fought so hard to block the original site in Grant Park.

Ward was 67 when the museum fight finally was settled, and his health was beginning to fail. He had never fully recovered from a broken arm and shoulder blade and was forced to give up the sports he loved—golf, horseback riding and driving his new Stanley Steamer. He grew weaker. He and his wife and daughter were preparing to go to California late in 1913 to escape the Chicago winter when he fell and broke his hip. It didn't heal properly. Six weeks after his fall, Ward developed pneumonia. He died a few days later, on December 7, 1913, when he was 69.

His contribution to the city was just beginning to be recognized at the time of his death. "Lake Watchdog Dies," said the *Tribune* headline on his obituary. City leaders made appropriate statements of respect, noting that if it hadn't been for A. Montgomery Ward, Grant Park would be a long line of buildings. They didn't mention, of course, that they were the ones who had wanted to put the buildings there. But the real impact of what Ward had done, and why, wasn't fully understood in 1913; and perhaps it is not yet fully understood and appreciated today. Plans for buildings in parks, wider roads in parks, more concrete in parks—they are as prevalent today as in 1890.

But even in Ward's lifetime, some people did understand. A man named J. J. Wallace spoke for all of them with a letter printed in the *Tribune*: "Who shall set a value on his service? The present generation, I believe, hardly appreciates what has been given them, but those who come later, as they avail themselves of the breathing spot, will realize it."

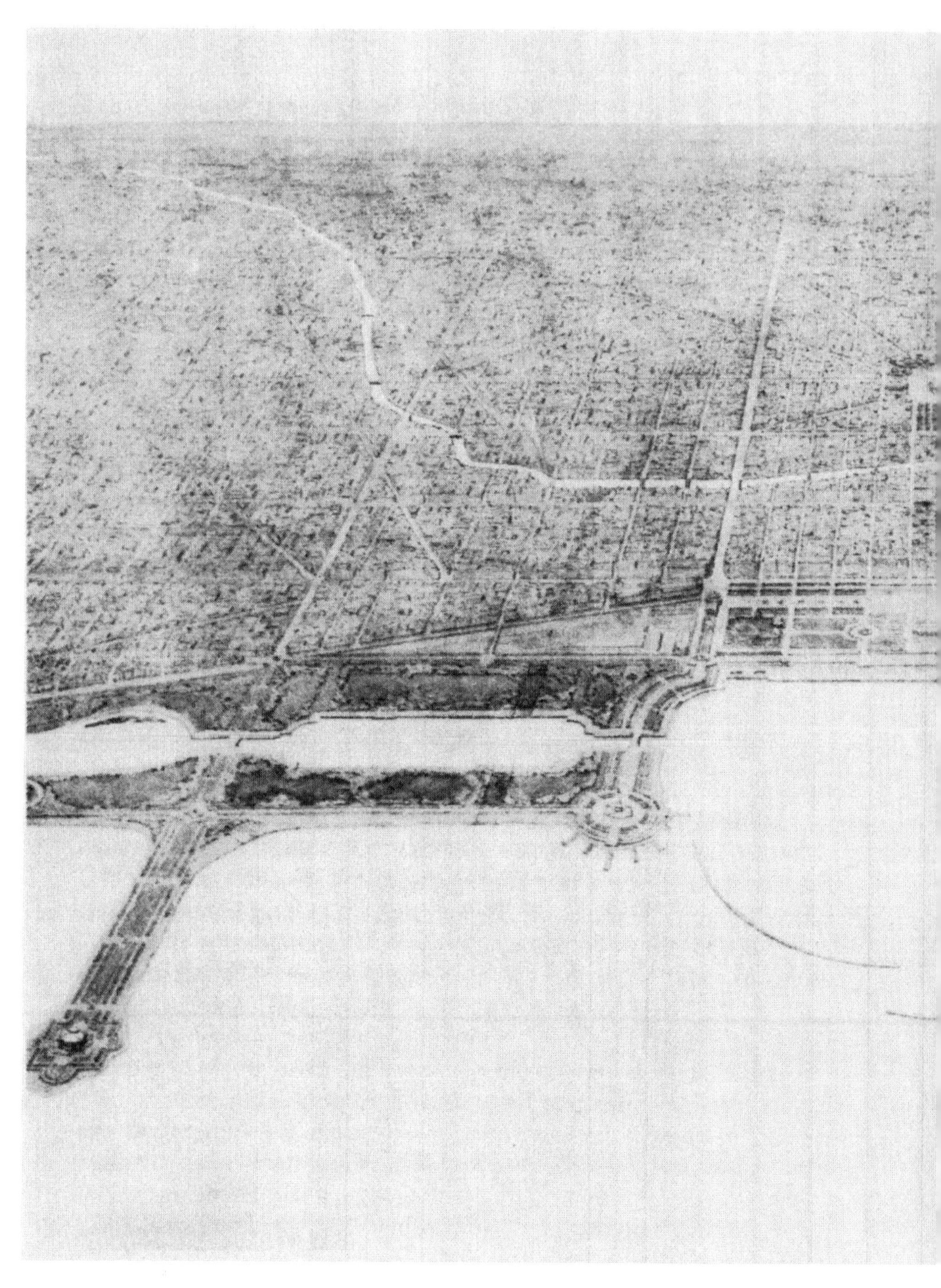

21. Burnham's 1909 Plan of Chicago, "Plate LXXXVII, view looking west over the city," showing a proposed civic center at Congress Street, Grant Park, and symmetrical harbors. Courtesy of the Chicago Historical Society. ICHi-17138

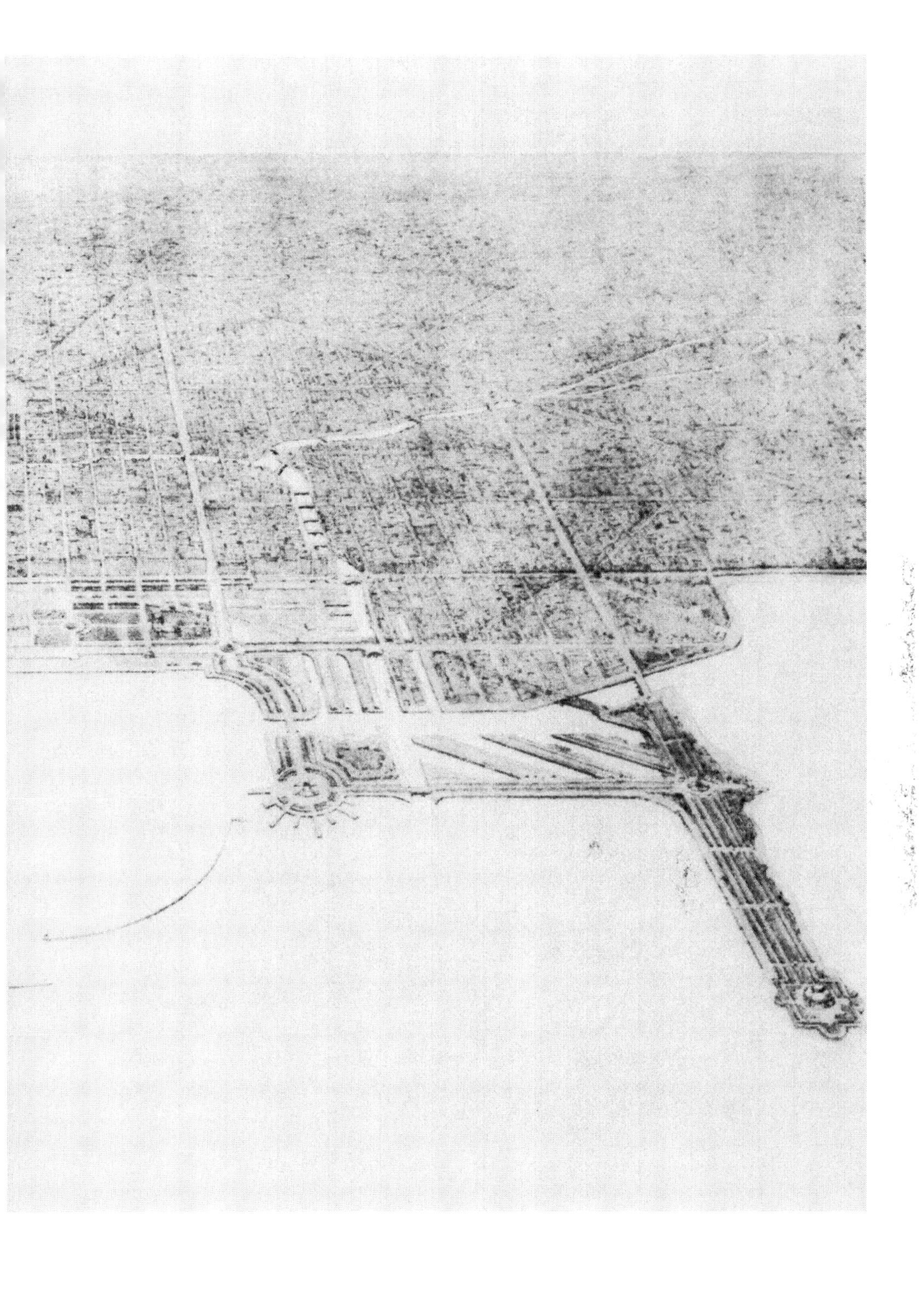

22. Grant Park's Buckingham Fountain in 1928, a year after its dedication. Kate Buckingham gave it to the city in memory of her brother, Charles. Detail of photo by Kaufmann & Fabry, courtesy of the Chicago Historical Society. ICHi-20500

23. The 1933 Century of Progress Exposition in Burnham Park, photographed by the Chicago Aerial Survey Company. Courtesy of the Chicago Historical Society. ICHi-0241

Chapter 8

Daniel Burnham Makes a Plan

Daniel Burnham, large, genial, energetic offspring of New England Puritan stock, was as talented at promotion as he was at architecture. Add to that his gift for creating stirring phrases, and you have the ideal man to change a city.

Without Burnham, and his particular combination of talents, Chicago and a number of other metropolitan areas would be vastly different today. He cherished beauty and he worshipped order. That's why he could plan steel-frame skyscrapers, clean and efficient, and also produce the extravagantly beautiful but wildly unrealistic White City of 1893.

"Beauty has always paid better than any other commodity," he told Chicago business leaders in one of his dozens of after-dinner speeches. "Noises, ugly sights, ill smells, dirty streets tend to lower average efficiency. In a state of good order, every businessman in Chicago would make more money than he does now." (He knew how to touch the heart of a tycoon.)

"Make no little plans," he advised. "They have no magic to

stir men's blood . . . make big plans, aim high in hope and work, remembering that a noble, logical diagram once recorded will never die, but long after we are gone will be a living thing asserting itself with growing intensity."

And: "Let your watchword be order, and your beacon beauty."

For a man with these convictions, Chicago in the mid-1890s must have been intolerable. It was a bursting, madly growing factory city of well over 1,000,000, a mass of stockyards and slums, skyscrapers and class strife, poets and boodling politicians. Its workingmen suffered in miserable poverty, and every so often their rage boiled out in bloody battles with police or with their bosses' anti-union squads. Probably the only well-organized neighborhood in the city was the array of gambling houses, saloons, dance halls, bordellos, voodoo doctors and penny arcades called the Levee. It was so famous that when the Kaiser's brother Prince Henry came to Chicago, he asked to be taken to the most elegant of the bordellos, the Everleigh Club at 2131 S. Dearborn. Madames Ada and Minna Everleigh tossed a lavish banquet in his honor, memorable because the men sipped champagne from the ladies' slippers. Recorders of such events say this had never been done before.

There were only a few automobiles in town, yet traffic was a mess. Wagons and pedestrians jammed up for interminable waits in front of the narrow drawbridge at Rush Street, the chief connection between the North and South Sides. The river banks were crammed with wharves and sheds. The West Side's busiest thoroughfare, 12th Street, ran narrow and cluttered between shacks and tenements. Michigan Avenue near the river was packed solid with old buildings—a narrow, dark, teeming quarter.

True, there were lovely lakefront parks on the North Side and South Side, and because of Montgomery Ward it looked as if the downtown shore would finally be cleaned up. But between these patches of green beauty were railroad tracks and freight yards, swamps and old mills and breweries.

So one day in 1894, soon after the Columbian Exposition closed, Daniel Burnham sat down and sketched what an ideal city should look like. Beauty and order, those were his inspirations. His firm by then was world famous, with an annual income greater than that of any other architectural company in the country. His partner, John Root, the intellectual artist-designer of the pair, had died shortly before the exposition opened; and Daniel Burnham, the mover and pusher of the company, the man who got things done, wanted to devote more time to the newborn profession of city planning. In the next 10 years he would create municipal plans for Washington, Cleveland, San Francisco, Pittsburgh and Manila—never charging a fee, because he was a rich man now and this was pleasure more than work, his gifts to the cities where his buildings rose.

His sketches of 1894 concentrated on the lakefront. He wanted to make it one long playground for the people, from the city's northern tip to the southern border. Inspired by Olmsted's landscaping of the south parks and the city planning concept of the Columbian Exposition, he sketched in lagoons and harbors formed by a string of man-made offshore islands. Running through it all was an "Outer Park Boulevard," meant to be a scenic curving drive, not a major arterial roadway.

Other people had been thinking about a link between Jackson Park and the newly emerging Grant Park. Among them were J. W. Ellsworth, the aggressive president of the South Park Commission, and leaders of the Commercial Club, an elite organization of the city's 60 top businessmen.

In 1896 the Commercial Club, at Ellsworth's urging, invited Burnham to discuss his lakefront plans at a dinner meeting. He was eloquent, as usual—a master salesman. Some of the club members he sold were the biggest names in the city: Marshall Field, Phillip D. Armour, George M. Pullman. Within a few months Burnham had shown his sketches and delivered his poetic pep talks to the Chicago Woman's Club, the Art Institute Board and the Merchants Club, an executives' group similar to the Commercial Club. To the Merchants Club, he said:

The lake has been singing to us many years until we have become responsive. We see the broad water, ruffled by the gentle breeze; upon its breast the glint of oars, the gleam of rosy sails, the outlines of swiftly gliding launches. We see racing-shells go by, urged onward by bronzed athletes. We hear the rippling of the waves, comingled with youthful laughter, and music swelling over the lagoon dies away under the low branches of the trees. A crescent moon swims in the western sky, shining faintly upon us in the deepening twilight. . . .

And what sort of prosperity is this which we should foster and maintain? Not that for rich people solely or principally, for they can take care of themselves and wander where they will in pursuit of happiness; but the prosperity of those who must have employment in order to live.

Besides, he told them, all this beauty for the workingman would be sound business. Adequate parks for recreation would offset the perils of overcrowding, disorder, vice. They were a "measure of precaution" against civic unrest. A labor force is efficient, he said, only if it is comfortable and content.

The Merchants Club was dazzled. The members asked Burnham to draw a complete city plan that would become the club's long-range project. The Commercial Club, equally inspired, had the same idea. In 1907 the two clubs merged under the Commercial Club's name, giving Burnham the support of most of the city's business and industrial establishment. Burnham would take no salary, but the club did give him $100,000 for a staff and expenses. He hired architect Edward H. Bennett to work on the drawings and designs while he concentrated on mapping out objectives and preparing the report.

To lay the legal groundwork, members of the Commercial Club went to Springfield to lobby for a bill giving the South Park Commission the right to obtain the 5⅓ miles of shore between Jackson Park and Grant Park through condemnation proceedings. But somebody else was lobbying down in the state capital. The Illinois Central Railroad managed to get its property excluded from the bill—which pretty well made the measure worthless. The new act did permit the park commissioners to bargain for the land and the riparian rights, but they were unable to reach an agreement with the railroad.

Mayor Fred A. Busse, the rugged, rough-talking, quick-thinking product of a tough North Side neighborhood, was also caught up in Burnham's dream. He didn't seem the type, but he said he was inspired to build "a City Beautiful." He named a high-powered Lake Shore Reclamation Commission to help the park commissioners in their negotiations with the railroad. The new group filed suit to recover all lakeshore land that had been filled by private owners. This property, the commission maintained, legally belonged to the state of Illinois. The lawsuit dragged on, but it had the effect the commission wanted: The railroad began to make concessions. It offered to give up its shoreline rights between 12th Street and 51st Street, the point at which the tracks turn away from the shore, and to cover its unsightly mess of tracks between Monroe Street and 12th Street. In return, it wanted to increase its right-of-way by 160 acres and acquire an additional 20,000 square feet of land between 12th and 13th Streets for a new terminal.

The commissioners felt they had a good bargain, but Alderman Charles E. Merriam, the distinguished University of Chicago political science professor, thought the Illinois Central should give a lot more. In particular, he wanted the railroad to electrify its line, thus removing the polluting coal smoke from the lakefront, and to depress its tracks along the shore so the trains wouldn't be so noticeable. He led a successful campaign in the City Council against the proposed settlement, and the commissioners and the railroad went back to the bargaining table.

In the meantime, Burnham and Bennett had finished their Chicago Plan. It was published in July, 1909, with magnificent color illustrations by Jules Guerin and an introduction by Burnham at his best: "First in importance [to the city] is the shore of Lake Michigan. It should be treated as park space to the greatest possible extent. The lakefront by right belongs to the people . . . . Not a foot of its shores should be appropriated to the exclusion of the people . . . ."

He envisioned a green belt of shoreline playground running for mile after mile, all the way north to Wilmette. Chicago's abundant supply of waste and dredgings was to be dumped into the lake to create a belt of islands—four between North Avenue and Touhy Avenue, one off Dempster Street in Evanston, and a peninsula and an island off the Wilmette shore. From 12th Street to 55th Street a belt of islands, connected by bridges and pedestrian pathways, would form a lovely lagoon about 700 feet wide and more than five miles long. The islands would have children's playgrounds, bicycle paths, bathing beaches, flower gardens, restaurants. On the shore side of the lagoon between 12th and 22nd Streets he wanted "a great meadow developed as an athletic field, with a central gymnasium, outdoor exercising grounds, beaches."

Grant Park would be the formal focal point, "the intellectual center of Chicago." He envisioned a complex of museums and libraries here. (Montgomery Ward changed all that.) The symmetry was exquisite, a joy to see in Guerin's paintings. At the park's north end a beautiful circular harbor curved into the lake, the mate to the circular park curving out at 12th Street. Between the two was an arc of shoreline, with small craft harbors. Long peninsulas of parks and walks jutted into the lake at Chicago Avenue and at 22nd Street. The one at Chicago Avenue included a series of docks for package freight steamers. Already, it was assumed the heavy cargo would come to Calumet Harbor.

The rest of the city would also have order, harmony, beauty: A highway system, mainly following railroad tracks; a plan for consolidating freight lines into a central depot, outside the downtown area; huge parks of forests and streams on the city's outer fringes; a broader 12th Street and a diagonal avenue to give the West Side good access to the central city; a wider Michigan Avenue, with a double-deck bridge connecting it to the North Side, and removal of the congested, obsolete industrial mess along the river mouth; a double-deck road curving along the river bank; a civic and cultural center at Halsted and Congress Streets.

It was the prototype for twentieth-century city planning, viewing the metropolis as one interrelated organism: efficient, logical and neat. To this day, all city plans flow from it to some extent.

The newspapers loved it and for weeks were filled with stories and pictures of the Chicago Plan. It would, said the *Chicago Examiner*, create "the most magnificent protected waterway for the pleasure of the people ever realized in any part of the world." Guerin's illustrations went on display at the Art Institute and were the top attraction. And, giving the Plan the biggest possible push, was a promotional genius: Charles H. Wacker. He was 53 years old in 1909, and had successfully run his family's brewery. His great passions were German music and making friends. He was a ruddy, cheerful, optimistic man, a master organizer—the quintessential civic booster.

The Commercial Club gave the plan to the City Council as a gift. Mayor Busse twisted a lot of arms belonging to his "lowbrow crowd" of aldermen, as he called them, and appointed a Chicago Plan Commission to map out steps for implementing the great work. Wacker, who had been a leader of the Commercial Club's plan committee, was named chairman. As his chief of staff he chose Walter D. Moody, a forerunner of today's public relations experts. (Moody referred to it as "the field of scientific promotion.") The two were so effective that just about everyone in Chicago who could read eventually knew a lot about the Chicago Plan. Children studied it in school; simplified, shortened versions were mailed in booklet form to every property owner; films were distributed; and sooner or later every club and church group was asked if it would like to hear a speech—illustrated with charts and pictures—on the Chicago Plan. After 60 years, Wacker's campaign still stands as the most energetic public information effort in the city's history.

It got results. The City Council approved plans to widen 12th Street, later named Roosevelt Road, and to create the diagonal Ogden Avenue and the double-deck Wacker Drive along the river. A bond issue of $3,800,000 was passed to widen Michigan Avenue and build the double-deck bridge to the street's north branch.

There was so much enthusiasm for the plan that the Illinois Central Railroad realized it could no longer count on the State Legislature to protect its shoreline rights. Pressed by the tough bargaining stance of Alderman Merriam and two South Park commissioners, John Barton Payne and Charles L. Hutchinson, it agreed to depress its tracks and electrify its line.

This agreement was reached in July, 1912. But the man behind it, the man who had inspired the city, never knew about this first great step. Daniel Burnham died in June while traveling in Germany. He was 66.

Before the agreement could be ratified by the City Council, it had to be approved by the federal government because it affected harbor development. Federal officials had been watching the negotiations nervously, and Colonel W. H. Bixby of the U. S. Corps of Engineers had warned the city that "to spend large sums of money in beautifying a city by turning its lakefront into a park is an enterprise commendable in itself, but to do so at a sacrifice of the harbor development of the city . . . is killing the goose which lays the golden egg."

To win federal approval of its park plan, the city agreed to build Municipal Pier, later called Navy Pier, north of the river mouth. It jutted 3,000 feet into the lake, cost $4,500,000, and was designed to handle both passenger and freight ships. In its rounded eastern end there was a huge dance hall, a theater and a restaurant. But, by the time it was begun in 1914, Calumet Harbor had snared the bulk cargo and railroads were carrying the smaller freight. Passenger vessels were about to lose virtually all of their business to the automobile. Not many years after it opened, the great pier was a fossil. Even its restaurant failed.

Once the harbor development was approved by federal

officials, the City Council was ready to move on the new lakefront ordinance. These were its major provisions: Build Daniel Burnham's five park-islands between Grant Park and Jackson Park, forming the long lagoon and bathing beaches to accommodate 120,000; fill in the shore parallel to the islands to create new parkland; and construct a lakefront sports stadium seating 100,000 people—the city's distorted version of the "great meadow" with free public sports facilities Burnham had recommended.

The Illinois Central agreed to depress its tracks from nine to 14 feet below ground level, construct 12 viaducts and new freight facilities, electrify 405 miles of its tracks and tear down its ugly red station blighting the lakefront at 12th Street. The station was to be replaced with a terminal conforming to the design of the new Field Museum.

The City Council approved the ordinance overwhelmingly on July 21, 1919. "This is the greatest day, barring none, in Chicago's history," exclaimed Charles Wacker after their vote. "It marks a new era. It is the beginning of the making of Chicago. It will be remembered later that it was this day which gave the city the most beautiful waterfront in the world."

He had reason to be jubilant. But the *Chicago Tribune* cautioned: "The meticulous historian will probably want to see results before admitting that Mr. Wacker's assertion is devoid of exaggeration."

The *Tribune* was right, of course. Neither side lived up to the terms. Both the city and the railroad made good starts, but then the Great Depression struck. And then World War II. And then . . . who cared about the lakefront ordinance? On balance, the railroad was much more faithful to the terms of the agreement than the city. It did lower its tracks, electrify, build the viaducts and the freight terminals. But the new station, harmonizing with the museum, never appeared. It is on the drawing board, though, as part of the new construction on Illinois Central air rights. Which is more than the city can say for that grand lagoon with its necklace of park-islands.

Seven months after the ordinance was passed, Chicago voters, steamed up by Wacker, approved a $20,000,000 bond issue to pay for the city's side of the agreement. The shoreline between Grant and Jackson Parks was filled in and named Burnham Park. The first of the five islands was completed and christened Northerly. And the South Park Commission selected the architectural firm of Holabird and Roche to design the sports stadium. The architects created a combination of a European soccer field and a Greek theater, built on an axis with the Field Museum and crowned by Doric colonnades. John Barton Payne, president of the South Park Commission, thought it was elegant. But Jimmy Callahan, former manager of the Chicago White Sox, saw the plans and thought they were foolish. He went to Edward J. Kelly, a rising politician who was chief engineer of the Sanitary District and a member of the park commission. You can't play baseball on that field, Callahan told Kelly; it's only 300 feet wide. You can't see a football game on that field; the stands slope too gradually, and only a fourth of the seats will be between the goal lines.

He urged the park commissioners to widen the playing area and build steeper, double-deck stands. But the architects were outraged. Widen it? Then it would no longer conform to the width of the museum. Double-deck the stands? That would destroy the lovely Doric colonnades. Payne, strong man on the board, sided with the architects.

So the peculiar structure that is Soldier Field rose along the lakefront between 14th Street and 16th Street, at a cost of nearly $8,000,000. In 1927, before it officially opened, 145,000 people crowded in to see (or at least be near) Gene Tunney and Jack Dempsey during their famous "long-count" fight. After that, it was downhill all the way for the great Chicago Hippodrome. It was used sporadically for stock car races and soccer games, and for a few major football games so popular that fans didn't mind sitting where they could barely see the players. But most of the time it was empty.

Probably the most interesting bit of news concerning Soldier

Field occurred in 1950, when Ed Kelly's brother Evan, director of special services for the park board, was sued for divorce. His wife accused him of maintaining a love nest under the stands.

In 1970 park engineers warned that water seeping through the structure had made it unsafe. The alternatives were costly repair or demolition. Lakefront conservationists began to organize a tear-it-down campaign, but Mayor Richard J. Daley had other plans. He agreed to demolish Soldier Field but only if he could build a new $55,000,000 stadium on the lakeshore southeast of the field. It would be primarily for a professional football team, the Chicago Bears. An outraged public didn't want to spend that kind of money for the Bears, much as they loved them, so the mayor switched plans. He offered to renovate Soldier Field instead. People who had fought the new stadium apparently were satisfied with their half-victory, and there was little adverse reaction to the mayor's new proposal. So it seemed inevitable that the Hippodrome on the lakefront, with its sprawling 41 acres of parking lots, would be around for many years.

The construction of Soldier Field was a gross blunder, one of a series of departures from the Burnham Plan made in the 1920s in the name of expediency and short-term razzle dazzle. Another was the Monroe Street parking lot in Grant Park. The South Park Commission, busy filling in the shoreline during the mid-1920s, halted its landscaping of Grant Park at Monroe Street. Downtown office workers began to park their cars on the undeveloped park north of Monroe. Eventually, cars crawled all the way up to Randolph Street. No Montgomery Ward was around to complain, so instead of ejecting the autos or building an underground parking area covered with lawn, the park commissioners sanctioned the takeover. They built a huge, three-block-long, open air parking lot for 2,700 cars, usurping the entire northern portion of the park.

Private clubs began to appear in the public parks, given long-term, low-cost leases by park commissioners. A gun club and a tennis club, with initiation fees as high as $500 and annual dues of $200, opened in Lincoln Park.

In 1944 an appellate court ruled that Chicago parks were set aside for the equal benefit of all people, and that the Lincoln Park Gun Club had no exclusive right to occupy park land. This opened the way for park officials to evict the club, if they so desired. They did not. But the clubs did agree to permit nonmembers to use their facilities—for a fee.

Eight private yacht clubs were built, six of them with park officials' consent. The other two, the Chicago Yacht Club and the Columbia Yacht Club, were constructed in the water beyond Grant Park. Therefore, according to a State Supreme Court ruling, they were outside the park board's jurisdiction. After complaints in 1944 that the Chicago Yacht Club practiced discrimination in admitting members, president Charles Braun denied the club had restrictions, "except that we do not admit Jews."

After the appellate court ruling in the Lincoln Park Gun Club case, park officials ordered the yacht clubs on park property to give "facility" memberships to all applicants who owned boats and could pay the dues. But "social" memberships—the right to attend club parties and compete in club races—were still granted at members' discretion.

Spurred by the persistent criticism of independent Alderman Leon M. Despres, who contended that restrictive clubs had no right to use park property, the park officials inserted a clause in the club leases prohibiting discrimination because of religion or race. In 1962, when two prominent black boat owners were refused social memberships in the Burnham Yacht Club, there was an uproar in the press. The park board voted to cancel the club's lease, so club members quickly decided to accept the two. "It was the only way we could save the club," complained one member.

Not all park developments of the mid-1920s were discouraging. Lincoln Park steadily advanced north, buying up riparian rights and filling in the lake. By 1925 it reached Montrose

Avenue, and acquired Belmont Harbor and Waveland Fieldhouse with its picturesque carillon tower. The next leap, taking the park up to Foster Avenue, created Montrose Harbor and the huge Montrose-Wilson beach—a half mile long and 26 acres of sand. The two extensions gave the city more than 600 acres of parkland.

Grant Park got a magnificent $500,000 fountain, donated by Kate Buckingham in memory of her brother Charles, a benefactor of the Art Institute. Following a sketch from the Burnham Plan, it was set at the foot of Congress Street in a garden 600 feet square, with four smaller fountains in each corner. Burnham envisioned the park as a majestic Versailles-type garden, not an informal playground. The fountain brought the tone he wanted, and has drawn millions to the park. Two magnificent panels with thousands of rose plants extend for 317 feet on either side of the fountain. In the Versailles style, they are bordered with trees and alternate with panels of grass. Another very formal, beautifully arranged Grant Park garden lies just west of the fountain, with a stunning background of Chicago skyline that makes it a favorite luncheon picnic spot for downtown office workers and tourists. Called "Court of Presidents," it was designed for statuary; but so far only St. Gauden's "Seated Lincoln" has been placed there.

In Jackson Park, Mrs. Albion Headburg won a stay of execution for the old Palace of Fine Arts. Never meant to be a permanent structure, it had disintegrated badly since the Field collection moved out in 1923. The South Park Commission decided to demolish it. But Mrs. Headburg, a prominent club woman and charity worker, organized 6,000 women into a protest movement. Each contributed $1, and with that money, Mrs. Headburg had a small portion of the crumbling ruin repaired to remind people what it could look like. The cry "Save the Arts Palace" took hold and in 1924—thanks to Mrs. Headburg—voters approved a $5,000,000 bond issue to restore the classic structure.

Julius Rosenwald, president of Sears, Roebuck and Company

and a generous contributor to schools and housing programs for blacks, gave $3,000,000 for a new kind of museum in the renovated palace. He had never forgotten his eight-year-old son's fascination with an industrial museum in Munich, and so the famed Museum of Science and Industry was born. The only problem with the marvelous institution was that so many Chicago area families and busloads of visitors from neighboring states poured in to see the coal mine and watch themselves on color television and sample the marvels of mathematics and medicine and physics that the museum gobbled up more and more lawns for parking space. Underground parking was never seriously considered.

But throughout the active 1920s, the development that stirred the most interest and pride was the new island and the park strip emerging along the lakeshore. Chicago leaders were so pleased with this achievement that instead of moving on to Island No. 2, they planned a new civic venture: another world's fair.

The year 1933 would be the 100th anniversary of the chartering of the village of Chicago, and the business community wanted a gigantic celebration. The prime mover was Charles G. Dawes, the city's most distinguished citizen; he had served as vice president under Calvin Coolidge, and in 1925 was co-winner of the Nobel Peace Prize for arranging a plan for German war reparations. He came to Chicago from Washington to head the City National Bank and Trust Company. With his brother Rufus and James L. Simpson, chairman of the Chicago Plan Commission, he sold the city on the Century of Progress Exposition.

This time, selecting a site for the fair was no problem. It would be on the new Northerly Island, of course, and the landfill west of it.

But the Great Depression hit Chicago hard. City government was bankrupt by 1930 and banks refused to extend credit. Police, firemen and teachers went without pay. Yet plans for the fair moved ahead. Just the thing to snap the city back, the promoters said. They vowed it would be built without tax money, and they kept their promise by somehow managing to sell an initial issue of $10,000,000 in stock—in the midst of the Depression.

The theme was man's mastery over nature, and the motif was purely functional, totally efficient—a stark contrast to the classic splendor of the Columbian Exposition. A series of white buildings rose with geometrical precision along the shoreline, ready for the finest exhibits of science and industry assembled up to that time in the United States.

In February, 1933, Mayor Anton J. Cermak went to Florida to invite incoming President Franklin D. Roosevelt to open the fair. He was shot and killed by an assassin aiming at Roosevelt, and a stunned city temporarily halted its celebration plans.

But the fair opened as scheduled on May 27, with a spectacular feat. The lights were turned on with energy from rays of the star Arcturus, which were focused on photoelectric cells at a series of astronomical observatories and then transformed into electrical energy transmitted to Chicago. Arcturus was chosen because it was about 40 light years away—240 trillion miles—so the rays which lighted the fair had left the star about the time the Columbian Exposition opened in 1893.

The auto manufacturers' exhibits were always jammed, but so were the life-sized reproductions of dinosaurs and brontosaurus. Fair visitors saw the bathysphere that took William Beebe 2,200 feet beneath the surface of the sea, and the aluminum globe in which Auguste Picard soared 54,000 feet into the stratosphere.

But the most breathtaking of all was the Sky Ride. Two gaunt towers 64 stories high were connected at the 24th floor by cables along which rocket cars shot back and forth. Before the possibilities of air travel were recognized, it was believed cities of the future would be joined by devices like that.

The talk of the fair, though, was not science. It was nudity. A young woman named Sally Rand was dancing in an exhibit called the Streets of Paris, adorned only by two ostrich-feather fans. By mid-summer Sally was the most popular event at the fair.

Judge Joseph B. David was asked to issue an injunction ordering her to put on clothes. He refused. "Some people would like to put pants on horses," he said. Anyone offended by Miss Rand could walk out of the Streets of Paris exhibit, the judge noted, adding: "If you ask me, they are just a lot of boobs come to see a woman wiggle with a fan or without fig leaves. But we have the boobs and we have a right to cater to them. The Streets of Paris could starve to death for all of me. I go where there is a good glass of beer."

The mayor during fair days was Edward J. Kelly, the ex-park commissioner and Sanitary District boss who, with Democratic County Chairman Pat Nash, had taken control of the political machine built by Cermak. Kelly was not known to be a stickler for propriety, but when he visited the fair's Oriental Village he was shocked. The dancers were wearing such flimsy little things that the mayor was reported to have blushed and fled. He banned nudity from the fair. Sally Rand, in a huff, donned a transparent gauze costume and the Oriental dancers added an extra layer of sheer fabric.

In an exhibit bringing great French paintings to life, a young topless model, her hand strategically placed, represented Manet's "Olympia." Fair directors told her to hold a large bouquet of flowers. The exhibit manager complained that this insulted a great work of art, and he asked Judge David to issue an order restraining the directors from covering his model with flowers.

"I'll issue the injunction," the judge said, "but heaven help the lady if she moves her hand."

Presumably she didn't, and the fair thrived despite nationwide poverty. By the time it closed on November 12, 1933, more than 22,000,000 had seen it.

General manager Lenox R. Lohr, a former major in the Army Engineer Corps brought to Chicago by Charles Dawes to run the fair, announced it would be extended for another summer. But, he said, it would "swing to finer things"—reproductions of villages from around the world—to take people's minds and eyes away from Miss Rand and topless models. Nearly 16,000,000 saw it the second year.

The Century of Progress was such a stunning success that, as it closed, Mayor Kelly had a brilliant idea. Instead of all those parks and beaches and lagoons and flowers proposed by Daniel Burnham, why not give the people of Chicago a permanent fair? A year-round carnival, bigger and better than Coney Island. And, for the business community, a great convention hall, the finest in the world—and right on the lakefront.

24. The Field Museum, Shedd Aquarium, and Adler Planetarium in 1947. Burnham wanted a similar park and harbor development at the north end of Grant Park. Photo by Howard A. Wolf, courtesy of the Chicago Historical Society. ICHi-00940

25. The first McCormick Place, opened in 1960, was gutted by fire in 1967. The new McCormick Place, an improvement over the old concrete monolith, has glass walls, exposed roof trusses, and a plaza with a lake view. Photos courtesy of the Metropolitan Fair and Exposition Authority.

26. Citizens fighting for their park tied white bands around doomed trees in Jackson Park in 1965. Photo courtesy of Chicago Sun-Times.

27. Aftermath of slaughtered trees, removed to make room for the widening of Lake Shore Drive. Photo courtesy of Chicago Sun-Times.

28. View north toward downtown from Promontory Point, 1989. Photo by Juliana McCarthy, used with permission.

Chapter 9

The Age of Cement and Convenience

Like the Columbian Exposition before it, the Century of Progress managed to make money in the depths of a Depression. The moral, if any, was ignored by the men of 1893; but not by the men of 1934.

The 1893 Exposition had been coordinated with the development of Jackson Park and when it came down the city was left with a big, beautiful expanse of greenery and waterways. The Century of Progress went up on miles of raw, new landfill, undeveloped and unplanned. When it came down it left nothing but miles of raw, new landfill. But Mayor Ed Kelly and the fair's chiefs, Leonard Hicks, Lenox Lohr and Charles Peterson, had big plans for the area. If a fair could make money in a severe Depression, wouldn't it be good for the city to have one all year long, every year?

For Northerly Island, they proposed the "finest, most magnificent, most up-to-date recreation grounds in America, if possible the world," as Peterson put it. There would be a beach, a casino, and amusement-park funhouses, "but as far removed from obscenity or ballyhoo as possible," he said. On Burnham Park, across from the island, would be "a huge exhibition hall and convention hall, in a building some 1,200 by 500 feet."

Peterson, a businessman, had been appointed by Kelly to head the Mayor's Committee on the Burnham Plan, so his endorsement carried special weight. He was in the vanguard of a new breed of big businessmen; normally Republican, they would work closely with Democratic leaders in City Hall to further Chicago's commercial development. Unfortunately, their concern for the city's social and esthetic development was not always as pronounced.

Another trend in the Kelly era that was to become a Chicago tradition was the politicization of independent bodies like the park board, tying them closely to City Hall. By 1934, the park system had proliferated into 22 separate bodies operating big parks, little parks, playgrounds and boulevards. The largest of them—the South, West and Lincoln Park Commissions—had done outstanding work. But the system as a whole was inefficient and deeply in debt. The State Legislature consolidated it into one body, the Chicago Park District. Commissioners of the new board naturally had much more power than the members of the 22 little boards, so Kelly gave the jobs to men he could trust implicitly. His successors, Martin H. Kennelly and Richard J. Daley, continued the practice. Chicago's parks came under the control of powerful ward bosses, labor leaders loyal to City Hall, top real estate dealers and investment brokers who had supported and befriended the mayor. The early park commissioners had been familiar with landscape architects and city planners; the new ones were closer to ward committeemen and precinct captains.

One of the first acts of Kelly's new park board was the endorsement of the lake carnival-convention hall scheme.

The mayor said he was certain he could get a $25,000,000 loan from the federal Public Works Administration to build it, so the admission fee to Northerly Island, presumably a public park, would be a mere 25 cents.

"The project is the most momentous suggestion since the Chicago Plan was announced," he said. "What it may mean permanently to our city fairly staggers the imagination."

A number of Chicagoans had a pretty good idea what it would mean permanently to the city. The lakefront parks were a major source of civic pride and people were not about to sacrifice them.

The *Chicago Daily News* led the angry opposition, calling the plan "ballyhoo," and charging Kelly with trying to "smear the face of Chicago with a cheap and tawdry permanent fair, of a kind only the clever dullness of politicians could conceive."

Walter Fisher, who as a young attorney had been legal adviser to the Commercial Club's original Burnham Plan committee, organized architects and nature lovers and civic groups into a tough "Park and Lake Front Defense Committee." At a big public meeting in the Union League Club, the new group resolved that "no use of the public parks should be permitted which is not fundamentally recreational in character . . . no buildings should be permitted in the public parks if the payment of the cost of such buildings is primarily dependent on revenues to be derived from amusement enterprises, concessions, or leases to private interests for private profit."

Within a month, 550 groups had announced their opposition to the plan.

But Kelly succeeded in pushing through the State Legislature a bill providing for "exposition authorities" to run convention halls and fairs, and a bill permitting the Park District to lease up to 10 per cent of total park space to these new authorities.

Governor Henry Horner didn't like the bills, but neither did he want to offend Chicago's Democratic organization. So he permitted the bills to become law without his signature.

Kelly applied for the $25,000,000 federal loan, with a last-minute addition: an airport on Northerly Island. And, suddenly, the whole plan fell through. Harold Ickes, Secretary of the Department of the Interior, was strongly opposed to an airport on an island. It was too hazardous, he said. He refused to approve the loan. Without federal money, the penniless city government had no way to finance the project. So, temporarily at least, the lakefront was saved.

But what was saved? There was no money to landscape Burnham Park and Northerly Island, no money to develop beaches, no money to complete Daniel Burnham's lagoon-and-island chain. Money that was available was used to widen Lake Shore Drive, eating up acres of park space. The scenic park road had become a pioneer urban expressway. Lincoln Park continued to expand north during the Depression and immediate postwar years, when other park development was stymied, primarily because the drive—now the key north-south roadway—had to be lengthened.

As Lake Shore Drive grew, it virtually cut the city from its lakefront. This was particularly true in the stretch along Burnham Park, from 13th Street to 56th Street. For miles there was no road or pedestrian overpass leading to the park and its beaches, although the neighborhoods west of the park had become the city's worst slums, badly overcrowded and desperately in need of more recreation space.

World War II stopped development of Burnham Park and the islands that were to border it. It also stopped critically needed improvements in the mass transit system, in housing and slum clearance, in highway expansion to the suburbs. Living conditions worsened considerably after the war, when hundreds of thousands of new arrivals from the rural south crowded into the decaying central city. After 15 years of Depression and war and the mismanagement of Ed Kelly and his cronies, the city was staggering. Its slums were horrendous. Traffic barely moved on an obsolete road system. Equipment on the transit system was worn out. Racial wars flared as whites fought the spread of the black ghetto. The school system was infested with political hacks; no one became a principal without his ward committeeman's approval. Basic city services such as street sweeping and garbage pickup had fallen apart. Understandably, recreation and park development had no significance at all.

But the tragedy of the postwar years was not that the parks were ignored—that could eventually have been remedied. Instead, great tracts of them were destroyed, sacrificed to the gods Transportation, Commerce and Convenience.

The first major repudiation of the Burnham Plan and the city's commitment to preserve the lakefront for recreation came in 1946. The man behind it was aviation enthusiast Merrill C. Meigs, who had been publisher of the *Chicago Herald and Examiner* and the *Chicago American*, and was then vice-president of the newspapers' parent company, the Hearst Corporation. He had been a pilot since 1928 and for years had argued that owners of private planes should be able to land near their downtown offices. Midway, the municipal airport at the city's southwest edge, was too inconvenient, he said. Big businessmen might get so irritated they would move their operations out of Chicago.

Meigs had helped select a site for a second city airport, a huge Northwest Side tract that would become O'Hare. But that was even farther from the downtown area than Midway. He urged city officials to resurrect the 1935 plan for an airport on Northerly Island, designed principally for owners of private planes. The citizens groups that had fought the 1935 lakefront plans had faded away, and before new ones could organize the plan was approved by the City Council.

The five park commissioners agreed to lease Northerly Island to the city Department of Public Works for 50 years at $1 a year. Ground was broken for Meigs Field on June 20, 1947.

For the next decade, the same five commissioners would be just as willing to give away public parkland. All five were familiar faces in the city's political establishment. James H. Gately, a self-made man who quit school at 13 and built a profitable department store business, was Park District president from 1946 until 1967, when he retired at 84. He saw as his chief task the reduction of the Park District's huge debt. He did succeed in putting the system on a sound financial basis during his 23-year tenure as board president; but he didn't add much to the park system except several very profitable underground parking garages.

William L. McFetridge, president of the local Flat Janitors Union and a City Hall stalwart, was named commissioner in 1943 and served until his death in 1969. Mayor Daley replaced him with a good friend and staunch supporter, William A. Lee, president of the Chicago Federation of Labor.

Joseph W. Cremin, a real estate dealer, was appointed in 1946, soon after he resigned from the Board of Education during the scandals over political influence in the schools. When he died in 1956, Daley appointed John F. McGuane, a real estate dealer whom he described as a "lifelong friend." And when McGuane died four years later, Daley appointed Joseph L. Gill, Municipal Court clerk during the notorious bail bond and Traffic Court scandals of the 1950s. Gill was then 75.

John (Little Jack) Levin, whose West Side restaurant was famous for cheesecake and hosting the political gatherings of powerful ward boss Albert J. Horan, was a commissioner from 1946 until 1968, when he was 82.

But for two decades after World War II the strong man on the board was Jacob M. Arvey, who was appointed commissioner about the same time he took over the chairmanship of the Cook County Democratic Committee. Arvey was a shrewd political talent scout who had picked the late Adlai E. Stevenson II and former Illinois Senator Paul H. Douglas for high office. But his Park District interests seemed largely directed toward building a powerful army of political patronage workers.

The Civic Federation, an independent watchdog group that keeps track of Chicago's patronage system, reports that 2,000 of the 3,000 Park District workers who should be civil service employees are "temporary" employees—Chicago's euphemism for patronage workers. That means they are precinct captains, assistant precinct captains and others who have proven their worth on election day. "Every patronage job," Arvey once said, "brings in 60 votes."

For 20 years, these were the men charged with maintaining

and developing Chicago's world-famed park system. Their decisions were disastrous.

Meigs Field did become a busy little airport for executives, as its sponsor had predicted. But it was an extravagant taxpayers' gift to the businessmen, costing the city from $200,000 to $300,000 a year. Every time there were heavy fogs or strong winds over the lake—not at all unusual—Meigs had to close. Periodically, planes bound for Meigs would fall into the lake, buffeted by the unpredictable offshore winds. Pilots complained about the hazards of taking off from Meigs. When the 70-story Lake Point Tower was built two miles north of the airport, Leonard Kmiecek, secretary of the Chicago Area Pilots Association, said a plane taking off at 120 miles an hour from Meigs "could hit this building in one minute unless it makes a turn out into the lake." When visibility is poor, he said, "buildings loom up before a pilot can see them."

With the completion of an expressway to Midway Airport, the chief argument for Meigs Field—convenience—disappeared. It took only a few minutes longer to reach downtown Chicago from Midway than from Meigs. But there were several million dollars worth of buildings and runways invested in Meigs, and replacing costly concrete with open space is a revolutionary concept in city government. So Meigs is likely to remain on Daniel Burnham's lone island. Herbert H. Howell, chief of planning in the city Department of Aviation, said pilots had better learn to cope with their problem. "They will have to get along with high-rise buildings along the lakefront," he said to Kmiecek at a City Council hearing. "Eventually there will be a solid wall of them there."

As Meigs was nearing completion, the city's Department of Public Works selected sites for two huge new filtration plants, one for the North Side and one for the South Side. Both were on the lakefront. One would occupy 61 acres of landfill extending into the lake east of Ohio Street, and the other, almost as big, would jut into the lake east of 79th Street, at the southern border of Rainbow Park.

Nathaniel Owings, a widely respected architect and chairman of the Chicago Plan Commission, objected. There was no sound engineering reason for putting the plants on the lakefront, he said. They could go on severely blighted land; certainly Chicago had enough of that. But that would have added about $16,000,000 to the cost of the plants, estimated at close to $200,000,000. And, according to the report from city engineers, "Other things being equal, that project is the best which involves the least expenditure."

Other things were not "equal," Owings maintained. "We are opposed to anything that will detract from Chicago's magnificent lakefront," he said. "It is a prized asset that should be saved for recreational and cultural developments."

He argued that building the plants on the lake would be "a violation of a policy accepted by the Plan Commission since its origins, and a violation of a policy recognized in Grant Park by the Montgomery Ward decisions . . . . that is, that no construction be permitted along the lakefront unless it is built for the recreational and cultural use of the people."

To make its point clearly and strongly, the commission passed a resolution on October 4, 1948, that was to become a rallying cry for lakefront defenders:

Be it resolved by the Chicago Plan Commission that:

WHEREAS, Chicago's lakefront is of great natural beauty and offers the city one of its best opportunities for development of facilities for the recreation and health of its citizens; and

WHEREAS, The dedication of Chicago's lakefront to such facilities and its improvement and beautification have been an important part of the Chicago Plan since its inception, and hundreds of millions of dollars have been expended in achieving this program in large measure; and

WHEREAS, From time to time it is proposed that portions of the lakefront be used for purposes inconsistent with such development;

IT IS HEREBY DECLARED, That it long has been and shall remain the policy of the Commission to oppose in principle any use, public or private, of Chicago's lakefront (except such uses as may be requisite for harbor or terminal facilities for passenger and freight vessels between Grand Avenue and Randolph Street and for bulk and package cargo south of 79th Street) for other than recreational purposes.

But the Plan Commission, established by the City Council in Burnham's day to carry out its commitment to his plan and to pass on all public works, had no veto power. It refused to endorse the filtration plant sites, but the City Council approved them anyway.

The Park District commissioners quickly accepted the plans and gave away the shoreline.

To mollify the civic groups that had fought the plants, the city Department of Public Works promised that a big new park would be built adjacent to the North Side plant, with 16 tennis courts and a number of softball diamonds. A park was built, but it was only 10½ acres—little more than landscaping for the huge plant and the 70-story apartment building across the street. There was no room for tennis or softball, but there were pretty fountains and walks. On fine summer weekdays, young downtown office workers flock to the small green spot with their paper-bag lunches and Mexican carry-outs. The little park was named for Milton P. Olive III, an 18-year-old Chicago boy who was killed in Vietnam when he threw himself on a grenade to save his companions.

When he dedicated the little park and the plant, Mayor Daley observed: "People said it would spoil the lakefront . . . . But today these people of little faith have changed their minds; many now say they were wrong, and that the plant performs an important function . . . . a perfect example of good engineering and good judgment, and a model for the rest of the United States."

While the Plan Commission and its allies were fighting over

the filtration plant, Chicagoans were having fun on the shore of Lake Michigan around 23rd Street, the site of the Century of Progress Exposition. The Association of American Railroads brought a marvelous collection of old and new trains to the lakefront during the summers of 1948 and 1949. Their "Railroad Fairs" were great hits and, to the amazement of the sponsors, netted $500,000.

No one was more impressed than Robert R. McCormick, World War I colonel in the field artillery, staunchly conservative Republican, despiser of the New Deal and, like his grandfather Joseph Medill, editor and publisher of the *Chicago Tribune*. He loved fairs. He never got over the fact that the Century of Progress made money in the Depression.

So, when the Association of Railroads decided not to extend its fair for another year, the *Tribune* put up $50,000 and collected another $950,000 from downtown stores, hotels, restaurants and other businesses to finance it. But, partly because of bad weather and partly because two summers of railroads had been enough, the third fair lost money.

Lenox Lohr, the president of the Museum of Science and Industry and manager of both the Century of Progress and the Railroad Fairs, told McCormick that a fourth fair would be disastrous. Even so, the colonel was determined to have a fair—any sort of fair—on the lakefront, the 23rd Street site in Burnham Park. He told W. Don Maxwell, his managing editor, to work it out somehow.

Maxwell knew where to start. Early in 1951 he assigned George Tagge, the *Tribune* reporter covering the State Legislature, to push through a bill creating a fund for "industrial, scientific, educational and cultural" fairs and exhibits, financed with a tax of one cent on each dollar received by the parimutuel race tracks. A similar race track tax financed agricultural fairs, and the *Tribune* argued quite logically that metropolitan areas should get help for big-city fairs.

But race track owners were afraid an extra tax would drive away business. Governor Adlai E. Stevenson II was caught in a fight between Colonel McCormick, publisher of the *Tribune,* and John S. Knight, publisher of the *Chicago Daily News.* Knight had an interest in a race track and did not want the tax. At the urging of prominent Chicago merchant Joel Goldblatt and Richard J. Daley, then Cook County clerk and Stevenson's former director of revenue, the governor let the bill become law without his signature.

Maxwell and James L. Palmer, president of Marshall Field and Company, traveled to Toronto to look at its city-run fair. They were particularly impressed by its exhibition buildings and decided that an exhibition hall for trade shows and conventions would be even more beneficial to the city than a fair for summer tourists. "That's the first thing to shout for," Palmer told Maxwell.

An exhibition hall required new legislation. So in 1953 Tagge pushed through another *Tribune* bill, this one to permit construction of the hall with money from the industrial fair fund set up two years earlier. The bill angered William Wood-Prince, manager of the International Amphitheatre, largest of the city's privately owned convention halls. It was peculiar, he said, that the *Chicago Tribune*, foe of government intervention in business affairs, should try to drive out private enterprise with such a socialistic scheme as a tax-supported convention hall. After that, the *Tribune* referred to Wood-Prince as "the Easterner who under terms of his uncle's will operates the Amphitheatre."

The *Tribune* had the support of the new Republican governor, William G. Stratton. It had enormous influence with Republicans in the Legislature. Also, quite a few Democrats were happy to do the "*Trib*" a favor and perhaps someday be spared its wrath.

So the bill passed, and everyone knew exactly the spot the *Tribune* had in mind for its exhibition hall: 23rd Street on the lakefront, site of the Century of Progress and Railroad Fairs so dearly loved by Colonel McCormick.

All the Chicago Plan Commission could do was issue a poignant "Statement of Principles":

Any project of this magnitude should be planned to afford the greatest possible recreational and educational benefit to the people of Chicago, as well as economic gain.

Since the basic character of the convention hall is commercial, the project should be located so as not to be in conflict with the [1948] Lake Front Policy. The location should be confined to a belt extending around the periphery of the Central Business District . . . the hall should be readily accessible and not in conflict with movement of traffic generated by the normal function of the Central Business District. The location should be in close proximity to both expressway system and transit system; emphasis should be placed upon sound estimate of parking facilities, both in conjunction with the project and for overflow; there should be ready access by pedestrians to "promenade" streets such as Michigan Avenue and Wacker Drive.

Strong consideration should be given to the redevelopment needs in the Central Business District, such as the replacement of obsolete structures, the stabilization of declining areas, or the achievement of such projects as riverfront beautification.

These guidelines could hardly apply to 23rd Street. It had no obsolete structures that had to be replaced—only free, open shoreline and park. It was far from the city's public transit system; the closest station was a mile away. The only road leading to it was Lake Shore Drive, crucial to traffic flow to the downtown business area and always jammed during rush hours. Parking lots would have to spread along the lakefront. There was no "promenade" street nearby; there was no street at all nearby except the busy Lake Shore Drive. And, of course, placing the hall on that site would violate the Plan Commission's 1948 policy statement preserving the lakefront for recreation.

A prestigious group of civic leaders joined forces to fight a lakefront hall, including the Chicago chapter of the American Institute of Architects and the Chicago Urban League, a group founded to help new black residents. The Urban League pointed out that Burnham Park stretched alongside a badly congested black slum; the park should be expanded the way the city had promised in 1919, the League said, not usurped for a project to benefit business and industry.

The Metropolitan Housing and Planning Council, a civic organization formed in the 1930s to fight slums and blight, organized the opposition into a solid front. Its members were men and women outstanding in the fields of planning, architecture, recreation, the arts, social welfare. The Chicago Real Estate Board and the Chicago Association of Commerce and Industry announced their opposition to a lakefront site, declaring that a convention hall should be close to mass transit and within walking distance of the major downtown hotels.

William Spencer, head of the Chicago Plan Commission, appealed to the park commissioners to speak out, to fight any attempt to grab the land entrusted to them. They were silent.

Early in 1955 the backers of the convention hall—owners of downtown stores and restaurants, airline executives and taxicab company heads—unveiled plans for a huge exhibition hall, 250 feet wide and 1,200 feet long, with 306,000 square feet of exhibit space on its main floor. The cost: about $35,000,000. The site: the lakefront's Burnham Park, off 23rd Street. Only about $10,000,000 would be available from the new race track tax, so that meant $25,000,000 would have to be raised some other way. A public authority would have to be created and empowered to issue tax-free revenue bonds. It was doubtful that voters in a referendum would approve a bond issue for a convention hall, so the *Tribune* executives proposed an alternative: Create a Metropolitan Fair and Exposition Authority as a municipal corporation, give it power to issue revenue bonds and enable the state and other political bodies to invest in the bonds. This would avoid the necessity of a referendum. Also, the Chicago Park District would need legislative authority to lease Burnham Park for the structure. Bills were prepared, and for the third time George Tagge went to Springfield to see them through.

He expected an easy time, since the project was endorsed by both Republican Governor Stratton and Richard J. Daley, who had become chairman of the powerful Cook County Democratic Committee in 1953 and was elected mayor of Chicago in April, 1955.

But as the legislative session advanced, Tagge noticed a surprising lack of enthusiasm among Cook County Democrats. Finally one of them told him why: The bills provided that the governor appoint six directors to the new Authority and the mayor of Chicago only five. Tagge asked Stratton if the bills could be amended to give Daley an equal number of appointees. Stratton was irked, since it would be the state that would have to buy the $25,000,000 in convention hall bonds, not the city. But since the *Tribune* was so vitally concerned, Stratton agreed. After that, the bills had the wholehearted backing of the Cook County Democratic organization.

They passed, and in Springfield the giant hall became known as "Tagge's Temple."

Colonel McCormick died in March, 1955, and Maxwell was determined to make the hall a monument to him. He suggested appointments to the new Authority to both the governor and the mayor, and for the most part they went along with him. The presidents of several State Street department stores were named to the board; so was Arthur M. Wirtz, whose family had an interest in the Chicago Stadium, a sports arena; Otto K. Eitel, manager of the Bismarck Hotel; Henry Crown, one of the richest men in the United States, part owner of the Conrad Hilton Hotel and the Palmer House, and head of the firm that would supply construction materials for the hall; Patrick F. Sullivan, president of the Chicago Building Trades Council—labor's support would be essential to speed construction; Lenox R. Lohr, the veteran of the Century of Progress Exposition and the Railroad Fairs; and several steel and coal and railroad executives.

In March, 1956, the Authority asked the Park District to lease 34 lakefront acres around 23rd Street; 8 additional acres would be leased later. A long line of opponents arrived at the customary public hearing to speak against the lease. The site was too far from hotels and transportation, they said; it would create monumental traffic jams; it violated the city's commitment to preserve the lakefront for recreation.

But Park Commissioner Arvey said he didn't think a convention hall was a "commercial" enterprise, as the opponents charged. "It doesn't penetrate any particular group. It benefits all of our population."

If the park commissioners refused to grant the lease, he said, the Authority would need millions of dollars to buy a site and time to clear it, thus delaying the project by several years. And both the mayor and the governor had said the hall was needed right now.

One State Street merchant, arguing for the site, said he had visited the Burnham Park beach recently and had found "nothing but the breeding of juvenile delinquency."

The five commissioners voted unanimously to grant a 40-year, rent-free lease.

The Chicago Plan Commission had not approved the site, despite several appeals from Mayor Daley. The *Chicago Daily News* played up the commission's reluctance, but the mayor was not disturbed. He created a new city planning department, directly responsible to him. The old commission of citizen volunteers that had given birth to the Burnham Plan was retained, but its significance was greatly diminished. Chairman William Spencer resigned, stating that a planning body ought to be independent of political control and ought to have some authority.

As the Metropolitan Fair and Exposition Authority was about to sign the lease, one of its members objected. Arthur Wirtz said he had given a lot of thought to the matter, and was convinced that no knowledgeable, unbiased person would endorse the 23rd Street site. The hall should be close to the downtown area, he said. Mayor Daley, anxious to avoid a split in the Authority, arranged another public hearing. A number of planning experts testified, urging that the hall be built on the air rights over the Union Station railroad yards in the southwest section of the downtown area. For $6,000,000, they said, the Authority could have this spot—convenient for cars and buses and commuter trains, within walking distance to major hotels. And the hall would enhance the nearby Chicago River banks.

The citizens committee formed by the Metropolitan Housing and Planning Council presented an overwhelming amount of testimony against the lakefront site, based on transportation studies, parking demands and recreational needs. Public improvements, the council said, should replace slums and eyesores, not green spaces.

Ferd Kramer, a prominent real estate developer and president of the council, warned that the lakefront site would create intolerable traffic complications.

Only two people spoke for the lakefront site, a representative of State Street merchants and an official of the Chicago Convention Bureau. State Street merchants were divided on the issue, however; John Pirie, president of Carson Pirie Scott and Company, spoke forcefully for a downtown location.

The preponderance of testimony and evidence against the lakefront site meant little to the Metropolitan Fair and Exposition Authority. Once it had listened to all the speakers and fulfilled the mayor's wish for a public hearing, it went ahead with its original plan and signed the lease for Burnham Park.

It looked as though the lakefront defenders were finished. But one last hope remained: the courts, the recourse Montgomery Ward had used again and again to save Grant Park.

The basis for a suit rose out of an economic feasibility study the Authority had commissioned from Booz-Allen-Hamilton, nationally known management consultants. Their report said that no matter how elaborate and well-located a convention hall might be, it would not significantly increase the number of conventions coming to Chicago. Of the thousands of nationwide conventions and trade shows, only 48 required more than 25,000 square feet of exhibition space. The new hall would have 306,000. Whether or not these 48 came to Chicago depended upon a number of factors other than the hall, the consultants said; many of them rotated their conventions between cities.

The consultants concluded the Authority would barely break even on the hall each year—providing the race track subsidy remained high. If it fell off, the Authority might not be able to meet its obligations to its bond holders.

With this as a basis, real estate man Kellogg Fairbank, an ardent lakefront protector, sued to block the sale of convention hall bonds to the state on the grounds that this would not be a reasonable and prudent investment. His attorney, Calvin Sawyier, was active in the Metropolitan Housing and Planning Council and was strongly opposed to a lakefront hall. Sawyier argued that the state had agreed to buy the bonds under fraudulent terms, influenced by overly optimistic predictions of the hall's financial future. He also contended that the Park District had no right to dispose of the submerged lands that were part of the lease arrangement.

Sawyier did not expect to win in the Cook County courts, particularly since his case was assigned to Judge Thomas E. Kluczynski, who had close ties to City Hall. But he hoped that through an appeal to the State Supreme Court he could delay long enough until a new state treasurer took office, hopefully one who would not buy the bonds.

Cases like this usually proceed slowly through the court system, but for some unknown reason this one raced through at an astonishing speed. Sawyier had asked that trial be scheduled for May 1, 1958, to give him three months to prepare his case. The judge said the matter was too urgent, and set a March 17 date.

After several weeks of hearings, the judge ruled against Fairbank and Sawyier. Still, the election was only six months away; and normally it would take at least that long to get a decision from the State Supreme Court.

Suddenly, a mysterious intervening plaintiff appeared—hired

by the *Tribune*, according to Fairbank's backers. He asked for the earliest possible hearing before the State Supreme Court. Sawyier objected that his case was being sabotaged. Three months later, the court handed down a unanimous decision supporting the convention hall. The fight was over.

On September 17, 1958, Mayor Daley and Governor Stratton jointly broke ground for the new hall, and it was named McCormick Place.

From the start, it was a bad-luck palace. Few people liked the way it looked: an immense, pale, squat cement box, heavy and graceless. When it was used for public shows, traffic was stalled for miles on Lake Shore Drive, as opponents of the site had predicted. Its parking lots stretched for blocks along a barren shore. Fewer then 40 per cent of the shows in the hall were open to the public, and only 10 to 15 per cent of the meetings held there were public. And in 1962, two years after it opened, Chicago had fewer conventioneers than in 1955.

But, fortunately for the Authority, race track business boomed. The tax subsidy to the hall grew from $1,326,000 in 1962 to $2,086,000 in 1966. It was a good thing, too. As the management consultants had predicted, the hall would not have been able to meet its obligations to bondholders without a hefty race track subsidy. In 1966, its best year, the hall earned $754,705 over operating expenses; but its bond obligations were $1,671,950.

The traffic problem was so serious that $62,000,000 worth of enormous, double-deck, looping expressway interchanges were built over Lake Shore Drive so motorists could reach the hall.

The only good thing about the concrete spaghetti loops was that they partially hid the ugliness of the hall itself: a barren box likely to deface the lakefront for a century.

But on January 16, 1967, several hours after workmen finished installing the displays and booths of the big National Housewares Exhibit, some sort of electrical flash accomplished what lakefront defenders had jokingly threatened to do for years: It destroyed McCormick Place. Fire gutted the structure, leaving not much more than grotesquely bent steel girders and charred remnants of walls.

"It practically amounted to euthanasia," wrote *Chicago Sun-Times* architectural critic Rob Cuscaden.

A team of expert investigators appointed by Mayor Daley turned up some startling evidence. The main exhibition hall did not have enough sprinklers to give adequate fire protection; key fire hydrants had been frozen; the drapes and skirtings of the display booths were not required to be coated with fireproof materials; the structural steel girders that buckled in the blaze were not covered with masonry or a gypsum product for fire protection.

John Waner, a heating contractor and former federal housing official who was running for mayor against Daley, suddenly had a campaign issue. McCormick Place, worth $40,000,000, had burned to the ground and there was evidence of gross negligence. Had taxpayers' millions been squandered on cheap shortcuts? Had city fire inspectors checked it out? As is usual for a Republican mayoral candidate, Waner was receiving little attention from the newspapers and television stations. Now, at last, he had an issue they could rally around. But the day he ordered his staff to get busy on McCormick Place research, he had a visit from George Tagge.

"He said I should shut up about McCormick Place," Waner told *Daily News* columnist Mike Royko, author of the Daley biography *Boss*. "He said that if I started blasting away at it, it might embarrass him and the *Tribune*, so I had to shut up. I said the *Tribune* wasn't doing anything for me anyway. So Tagge said, yeah, but at least they weren't hurting me. So, I gave up."

As the Metropolitan Fair and Exposition Authority prepared plans for a new hall, much bigger and twice the price of the old one, there was a brief flurry of activity from the people who had fought against a lakefront site. Providence, or at least an electrical short circuit, has given us a chance to correct a monumental mistake, they said. Let's build a new hall on a better spot. But, like Waner, they soon realized that fighting McCormick Place-on-the-lake was hopeless. The Authority announced that $7,000,000 could be salvaged from the ruin—too large a sum to sacrifice, according to city officials.

So the lakefront groups turned their energies into improving the design of the hall. The *Chicago Daily News* and the *Chicago Sun-Times* joined them in insisting the new structure must somehow do a better job of sitting on the lakefront.

Mayor Daley said he, too, wanted a better hall. It would not take up more lakefront space than the old one, he said. If it expanded, it would move west, not north or south. The parking lots would be underground, and covered with grass. The trucks would not be seen as they loaded and unloaded—a major complaint against the old hall, which cluttered the lakefront with trucks every time it had a big show. And a little gift was thrown in from the Park District. The lake would be filled in east of the hall, creating a small park to link with parkland to the north and south.

Half the mayor's promises were carried through. The hall did expand west, right up to the edge of Lake Shore Drive. But it also expanded along the lakefront. The old hall was 345 feet wide and the new one was 610 feet; the old hall was 1,095 feet long and the new one was 1,360 feet. A covered parking lot was built on the south side, but open parking lots continued to deface the lakefront to the north. A gray wall hid the area where the trucks loaded and unloaded, but a huge open-air truck parking lot was developed on land west of Lake Shore Drive rented from the Illinois Central Railroad. An asphalt road taking trucks and cars from 31st Street had to be added east of the drive, chewing up more parkland.

A narrow green park, about seven acres, was built east of the hall along the lake. There were plans for a bicycle path that eventually would connect the north parks with the south parks—a token, at least, to the city's old commitment to the Burnham Plan.

The best feature of the new McCormick Place was the

building itself. With glass walls, exposed roof trusses and a see-through plaza to the lake, architect Gene Summers of C. F. Murphy Associates succeeded in giving a light, airy feeling to a gigantic structure. It was an enormous improvement over the old concrete monolith.

The Metropolitan Housing and Planning Council, which had led the spirited fight against the lakefront hall, praised the new design. Council leader Rex J. Bates called it "distinguished and graceful, despite its large scale and utilitarian functions."

"The beauty of the building, with its open view of the lake afforded by its glass walls and its canopied central plaza, will be a scenic asset," he said, "and a positive addition to the Chicago school of architecture."

But at a hearing on the new plans, he expressed a fear that had begun to overshadow the colossus on the lakefront. The Illinois Central Railroad had won court battles to sell the air over its tracks and right-of-way for commercial and residential development. Real estate developers were entranced with the possibilities.

But what would happen to Chicago's lakefront parks? Would they be cut off from the city by a wall of high-rises over the railroad tracks? Or would the city government demand firm control of the air-rights development and protect the people's parks?

A new lakefront battle was shaping up for the park defenders, and they trained rigorously for it during the skirmishes of the 1960s.

Chapter 10

"You Can Have Too Much Green Grass"

Sun splattered through the leafy trees in Jackson Park one September afternoon in 1965, shining on a scene that looked like an old-fashioned movie set—something from *Perils of Pauline.*

Two women flung themselves against tall oaks, their arms wrapped around the mighty trunks.

A group of men bearing powerful chain saws advanced toward them, deadly teeth glinting on the churning belts. All around was a field of mud and ruin. Bushes had been torn up and scattered about, trees had been chopped down, their stumps left naked and raw in the mud. Bulldozers had scooped up the lawn and dumped it into a lagoon clogged with dirt and grass and dying bushes.

Just a typical day in the park during summer, 1965. Instead of picnics on the grass, there was human flesh pitted against chain saws.

Day after day women and children stood steadfast between the trees and machines. At night there were protest rallies around the stumps. One day Jackson Park neighbors marched silently back and forth in front of Mayor Daley's home, waving branches from their fallen trees. Early every morning people turned out with banners and signs saying "Save These Trees" and pinned them on the surviving oaks and willows. Every afternoon city work crews tore the banners down.

Men and women who called themselves the "Daniel Burnham Committee" were arrested for littering and fined. They went back to their park and got arrested again. And every time the police took them away, the city crews with the chain saws advanced farther, slaughtering more trees, tearing up more lawns and shrubs.

It was a summer of rage and anguish—all because of traffic jams. Auto movement during rush hours had become something of a problem on South Lake Shore Drive, so the city planned to widen the road from 47th to 67th Streets, straighten its curves and run it straight through Jackson Park. The new eight-lane highway would bisect the park from east to west, south of the Museum of Science and Industry. Part of the quiet old lagoon would be filled in, so the road could run over it and down the middle of the park to 67th Street. At 57th Street a mammoth elevated highway interchange would tower over the shore and the lawns. World-famed Jackson Park, prized legacy of the 1893 Exposition, would be mutilated.

The frantic efforts of the Daniel Burnham Committee—the rallies, protest meetings, angry pronouncements and petitions signed by hundreds of neighborhood residents—did bring results. The city halted the roadwork after it had gone about half-way through the park. Mayor Daley set up an advisory committee to review Jackson Park development and included several prominent architects and urban affairs experts in the group. On the committee's recommendation, the city paid $30,000 to a nationally known landscape firm, Johnson, Johnson and Roy, of Ann Arbor, Michigan, to design a plan for a better road and a better park.

The following summer the firm produced its plan, and the

protesting groups were happy. The road would be six lanes wide, not eight. It would run along the park's east border, not down its center. It would be depressed, disguised with attractive park-like embankments. Instead of pedestrian tunnels under the road, there would be overpasses to the beach. One of the major flaws of the park—and the city's other lakefront parks, the firm said—was the lack of safe, esthetically pleasing and convenient pedestrian crosswalks to the beaches.

The firm recommended underground parking for the museum, which meant converting the parking lots that had crept around the building into wide expanses of lawn and gardens. Offshore islands were proposed. The South Side, the planners discovered, had grown much more rapidly than its beaches; more than half of the city's 3,500,000 people lived south of Madison Street, but the South Side had only a third of the total beach acreage.

The Jackson Park plan was so widely praised that a few months after its unveiling the landscape firm was hired jointly by the Park District and the city Department of Planning and Development to prepare a $100,000 master plan for the entire lakefront. The results, city officials said, would guide their moves for decades to come.

There was only one reason for cynicism: Nothing happened after the highly acclaimed Jackson Park plan was completed. Soon after the city released the study with great fanfare, it hired another firm (a local company that frequently received city business) to prepare a $50,000 "feasibility" study of the $30,000 study. This firm hiked the road back to an eight-lane freeway, elevated or at ground level, with pedestrian underpasses rather than overpasses. As in the original city plan, the auto was dominant. A stroll to the beach would mean passing through a tunnel instead of walking in the sun. The interchange at 57th Street was much more massive than in the Johnson, Johnson and Roy plan and was elevated about six feet. Depressing it, the firm said, would cost an extra $2,500,000.

Civic groups and political leaders who had hailed the Johnson, Johnson and Roy plan were outraged. They felt betrayed. Alderman Leon Despres called the new plan "a monstrosity, devastating to the entire city."

"It will destroy the tranquillity of the park," said Gunnar A. Peterson, director of the Open Lands Project.

The city resolved the conflict by doing nothing. All work in the park was indefinitely postponed. The half-finished road remained, cutting through to the center of the park and then tapering off into the old, narrower, winding lane.

There were no new offshore islands, no new beaches, no new landscaping. Even routine maintenance deteriorated. The Japanese gardens, a heritage from the 1893 Exposition, died for lack of care. One fall day a group of disgusted teenagers and adults staged a "rake-out" in the filthy Jackson Park lagoon. By noon, they had filled four two-ton trucks with sludge and debris.

If that was the result of the expert Johnson, Johnson and Roy plan for one park, would a master plan for the entire lakefront fare any better? Besides, by the time the master plan was commissioned late in 1966, the words "lakefront plan" and "park study" and "new shoreline proposals" already had become cynical jokes.

In the previous two years, the city had announced three different plans for expanded lakefront parks. None of them had been implemented, which was probably just as well. They were sketchy and vague, concentrating heavily on landfill peninsulas, which caused grave concern among ecologists worried about the dumping of debris into the lake for the fill and the changing of the natural shoreline.

In August, 1964, the city Department of Planning unveiled the first of the three plans in its long-awaited "Basic Policies for the Comprehensive Plan of Chicago," a document which was to set forth the rules of future city development. It stated that the city "strongly reaffirms the 1948 Chicago Plan Commission

Lakefront Resolution, which declared that the lakefront should be used for recreational and cultural purposes, except for the sections between Grand Avenue and Randolph Street and south of 79th Street."

(That wasn't exactly what the 1948 resolution said. It had made an exception of Grand Avenue to Randolph Street only for essential "harbor or terminal facilities for passenger and freight vessels." But luxury high-rise apartment buildings already were scheduled to go up east of Lake Shore Drive at the foot of Randolph Street and at Grand Avenue, so the omission was not exactly an oversight.)

The 1964 Basic Policies report noted that Chicago was far behind other major cities in recreational space, and expressed "substantial need for expansion of the city's lakefront facilities."

A year later, at a press conference to celebrate his first decade in office, Mayor Daley announced precisely what he planned to do for the lakefront. He promised a $156,000,000 program of recreational development, including four man-made peninsulas on the South Side, with a music bowl, a golf course and a big harbor. (His plan was unveiled in the middle of the spirited and highly publicized protests over the Jackson Park roadwork.)

But the Park District never had money like that and had made no plans to get it, either from federal sources or through bond issues. About half of its relatively small budget of $80,000,000 to $100,000,000 a year went for operations and maintenance, and another chunk was used to pay off old bond issues. Only about a fourth to a third of the budget was earmarked for capital improvements, which often turned out to be parking lots and garish street lights and expressway-type bridges over lovely old lagoons.

The costly peninsulas promised by the mayor in 1965 never appeared, but then no one really expected that they would. M. W. Newman, the *Chicago Daily News* expert on urban planning and architecture, noted that "Mayor Richard J. Daley, with an unhappy reputation as a tree-chopper, is also gaining a name for building lakefront peninsulas-in-the-sky."

Again, it was probably just as well they weren't built, because they were hastily conceived sketches by the Park District staff, designed to still the protests in the park. They related to no overall plan, and were not particularly distinguished works of landscape architecture.

In November, 1965, Daley unveiled another plan at a big City Hall press conference. This one would add 600 acres of man-made land to the shoreline within 10 years. After a month or so, hardly anyone remembered it.

Late in 1966 the Comprehensive Plan of Chicago, years in preparation, was ready for the public. It also promised great things for the lakefront. "The Lake Michigan shoreline in Chicago is a priceless natural and man-made asset for the entire region," it said. "It is the most important single recreational resource in the metropolitan area and has been a special concern to citizens, planners and public officials throughout the city's history." The "general character" of lakefront development should be "open landscape," the plan stated: *Lake Shore Drive, as a limited access parkway through the lakefront parks, should have special design criteria applied to it, as appropriate and necessary, to make it fully compatible with the primary recreational use of lakefront land.*

Because of the special nature of public interest in the lakefront, extensive design and development controls are required if the city is to assure that each public or private project is to make an optimum contribution to the development of the lakefront and adjacent districts.

A major area marked for landfill expansion in the plan was the Far North Side, north of the northern boundary of Lincoln Park. Property owners had built up to the shoreline in this area, leaving only street-end beaches for the public. New parks also were proposed between 39th and 55th Streets, where Burnham Park had been half-completed three decades earlier, and never finished. A third area that would get new parks and beaches was 71st to 79th Streets, north of the industrial and harbor complex.

The resolutions of the Comprehensive Plan were very nice, and the proposed new lakefront parks were badly needed. But by this time the civic groups concerned with park development had grown rather weary of City Hall's grand announcements. They did not expect the resolutions to be followed or the proposals to be realized. And they were correct.

So, when Johnson, Johnson and Roy was hired to prepare the final, absolute, overall master plan to supersede all other plans, the park groups held no premature celebrations.

But the Johnson plan, completed in stages from 1968 to 1970, was imaginative enough to stir considerable interest and some optimism among the urban planners, architects and conservationists who had fought for the parks. In addition to expanding and beautifying the parks, it stressed getting people to them—leaping that concrete barrier, Lake Shore Drive. The goal was to create a major recreational facility for the entire metropolitan area. In recent years, the planners concluded, the lakefront was used heavily by people who lived near it; but it was diminishing as a playground for residents of outlying areas. "We were trying basically to bring the parks back into a people-orientation," said William Johnson, partner in the firm.

The plan was ready late in 1970. But well over a year later, it was still under wraps in City Hall. "We think it's been tabled," Johnson said. "There has been a real slow-down in the pushing of park planning." The firm repeatedly called city officials to ask when the study would be released to the public. "But we were told there is no publishing date at this time," Johnson said. In City Hall, a spokesman for the Department of Planning and Development said only that it was "still being evaluated."

Yet, if construction of McCormick Place and the years that followed had proved anything, it was that a master plan for lakefront development was sorely needed. Daniel Burnham's 1909 plan had been shelved for so long, and the parks had developed so haphazardly in the intervening years, that it needed thorough revision and updating. But the 1960s brought such critical problems that park development and planning had little priority in City Hall, except when women and children were tying themselves to doomed trees.

Chicago, like every other great urban area in the nation, was in desperate trouble. Its school system was so inadequate, especially in the black ghettos, that great numbers of children fell farther behind in intelligence test scores as they advanced through the grades.

Schools in the white neighborhoods were not a great deal better. Middle-class white families fled to the suburbs in increasing numbers, often because they wanted better schools for their children.

Another reason they fled: They wanted more recreation space. Chicago had 3.16 acres of parkland recreation space for every 1,000 persons, the lowest ratio of any major city except for Washington, D.C. The National Recreation Association recommended 10 acres for every 1,000 persons. The park system that in the 1890s was considered one of the finest in the world had not grown with the population. The ratio of park-recreation space to population was nearly twice as high early in the century as it was in the 1960s.

The housing shortage for low- and moderate-income families was acute, particularly if the families happened to be black.

Health care for the poor was appalling. In the most impoverished neighborhoods, 45 of every 1,000 infants died before their first birthday—an infant mortality rate as high as in Ceylon and twice as high as in Chicago's middle-class neighborhoods.

The city's public transportation system, unable to find adequate financing, repeatedly raised its rates. And each time more people stopped taking buses and drove their cars to work and to stores. The city spent billions of local, state and federal tax money on new expressways in an attempt to keep up with auto traffic, but it was a crazy race, impossible to win. When the John F. Kennedy Expressway to the Northwest area opened in 1961, traffic on North Lake Shore Drive fell off 20 per cent. Seven years later it had risen to its old volume and eventually exceeded it. The same traffic pattern occurred on South Lake Shore Drive when the Dan Ryan Expressway to the South Side was opened.

But city officials persisted in trying to solve the problem by building new roads rather than a new, expanded public transit system. Rapid transit trains were installed along the median strips of expressways—a good idea but poorly planned. No nearby parking lots were provided to enable people in outlying areas to use the transit system with ease. And the new lines stopped several miles short of the growing industrial complex northwest of the city.

Chicago in the 1960s struggled with rapidly growing public aid rolls, with drug abuse among the young, with angry demands for better health facilities and better housing, with strangling traffic and air pollution, noise pollution and water pollution. Like other cities, it didn't have the money to handle the problems. But Chicago had another handicap.

Key city departments—Health, Building, Streets and Sanitation—were hampered by the political patronage system, which hung on in Chicago long after it had died in other big cities. The Cook County Democratic Committee controlled tens of thousands of city jobs, selecting employes for their ability to bring in votes rather than their ability to pass civil service tests. Contracts to professional consultants and to developers of urban renewal property regularly were awarded to firms that found favor in City Hall.

The Chicago Park District was one of the casualties of the system. It had a competent professional administrator in superintendent Thomas Barry, but he was saddled with the highest percentage of political patronage employes of any governmental unit, nearly 66 per cent of all those who should be covered by civil service, according to the independent watchdogs of the Civic Federation. An agency beholden to the political patronage system often has trouble attracting top people. This was true in the city Health Department, the Building Department, and the Park District.

A partner in a nationally known architectural firm, discouraged after trying to persuade Park District officials to upgrade their design concepts and their design staff, said bitterly:

They need high-calibre landscape architects. They don't have an in-house staff that's worth a hill of beans. They have payrollers functioning as engineers, architects and landscapers. Their work is despicable. When you think of the traditions of the early park landscapers . . . it makes you sick.

Rex Bates, representing the Metropolitan Housing and Planning Council, presented the dilemma succinctly at a 1968 meeting protesting the decline of park maintenance:

The plans produced by the Park District staff for specific parks have often been inferior, inappropriate and insensitive to their settings in style and design. We fear that the talent of the Park District staff has diminished in professional stature. It appears to us that there is too much emphasis on construction and engineering, and too little on design, artistic quality and esthetics.

But the decade of inferior, hit-or-miss park development did produce one encouraging trend. For that, Chicagoans could thank the giant on the lakefront, McCormick Place. The groups that mobilized to fight it in the mid-1950s formed a loose coalition, ready to swing into action whenever parklands were threatened. In the forefront were the Metropolitan Housing and Planning Council, the Chicago Chapter of the American Institute of Architects, the Committee for Chicago's Parks, the Open Lands Project, the Chicago Heritage Committee, the Daniel Burnham Committee and neighborhood groups such as the Lincoln Park Conservation Association and the Hyde Park-Kenwood Community Conference. If it had not been for these groups, concrete would cover considerably more areas of lakefront than it does today.

The Hyde Park-Kenwood group achieved a victory of sorts in

the mid-1950s when it caught the Park District about to lease Jackson Park's lovely Wooded Island, created for the 1893 Exposition, to the Army for an anti-aircraft installation. The protests were so strong that Wooded Island was spared. Instead, the Park District gave the Army another choice spot—Promontory Point at 55th Street, a favorite picnic grounds. More parkland to house the missile battalion and its radar station was leased at 61st Street. A second missile site and radar station was leased near Montrose Harbor in Lincoln Park. Altogether, the Army had more than 88 acres of lakefront land. Rocket launchers for the missiles, microwave towers and green barracks and ammunition buildings rose in the parks. Lawns and flowers and shrubs and trees were cleared and fences were erected.

And by the time the installations for the Nike missiles were completed, the Nike was considered useless by military authorities.

Representative Sidney Yates, whose North Side district included Lincoln Park, unearthed a 1959 report from the Senate Armed Services Committee which declared that "the Nike-Ajax, Nike-Hercules system is virtually obsolete, and the further expenditure of funds to expand the system, except in overseas areas where it might have tactical value, is unwarranted." New bombers could destroy a city from outside the Nike's range. Yet, despite this evidence and the repeated protests from political leaders such as Yates and Representative Abner J. Mikva, whose district included Jackson Park, the Park District renewed the leases. The Army had proved its case, according to Park Commissioner Jacob Arvey, and that was that. As in the past, the commissioners seemed too willing to hand out parkland. In Cleveland, for example, city officials forced the Army to build its anti-aircraft installations on filled-in lakeland and on private property, sparing the public parks and beaches.

Finally in spring, 1971, the Army announced it would remove the missiles—long after Nike bases on privately owned land had been shut down and a decade after they were generally acknowledged to be worthless.

Throughout the decade, citizen battles for the parks turned out like that: half-victories, to be celebrated only because things would have been so much worse if no one had fought; but never very satisfactory, because the public knew another concrete-spreading plan waited just around the corner.

In 1962 the city announced plans to build a $4,000,000 elevated highway interchange that would tower 15 feet above Oak Street beach, the Near North Side's popular summer playground. The plan was unbelievably bad. It would have destroyed the open lake vista and relegated the beach to the shadows of a mammoth concrete interchange.

A powerful and persistent opposition of neighborhood groups reinforced by the Metropolitan Housing and Planning Council persuaded city officials to change the plan. The interchange was placed underground. The expanse of concrete along the beach was greatly increased, but at least it did not block the lake view.

If the opposition groups thought their Oak Street victory would have some impact on future lakefront planning, they soon were disillusioned. Plans were announced for a new concrete expanse. The Park District was about to tear up two blocks of lawn in Grant Park for a huge music bowl that would seat 20,000. Like the Oak Street proposal, the plan was totally unsuitable. It was true that the small bandshell in the south end of the park, built in 1932, was inadequate. Concerts held there—classical, pop and rock—were popular. But the benches around the bandshell could seat only 700. It faced a fork in Lake Shore Drive and Columbus Drive at the foot of the park, a noisy spot, full of gas fumes. Yet there were times on fine summer evenings when 8,000 and 9,000 persons crowded on the grass around the bandshell to hear the concerts.

But a bowl with 20,000 seats? Half that many would have been sufficient, and far more esthetically pleasing. On the rare occasions when bigger crowds would attend, benches could be placed on the grass. And why usurp two blocks of lawn when the huge Monroe Street parking lot at the north end of the park could be covered, and the bowl built over it? That would get rid of the park's major eyesore, add to its green space, and provide an ideal spot for a music bowl.

An impressive number of civic groups objected to the Park District's plan. The area's nature lovers—the Illinois Audubon Society, the Sierra Club, the Garden Club of Illinois, the Chicago Ornithological Society, the Midwest Open Lands Association—joined the fight led by the Metropolitan Housing and Planning Council to save the Grant Park gardens.

Chicago Sun-Times reporter Ruth Moore, a specialist in urban planning, asked James H. Gately, the long-time Park District president, whether he might reconsider.

No, Gately said in the interview, he didn't think he would. And he didn't see anything wrong with tearing up a few blocks of lawn.

"You can have too much green grass," he said.

The opposition grew, so much so that Mayor Daley said it might be a good idea if the size and location of the music bowl were "reviewed."

In the end, the park was saved by an old nineteenth-century law providing that no structures could be built in Grant Park without the unanimous consent of the property owners facing the park. Three of the 46 owners refused to consent to the music bowl. Chicago Community Trust, representing the ownership of Orchestra Hall across from the park, offered to help underwrite the expenses of a master plan for Grant Park. Certainly one was needed. But the Park District and city officials let the offer die. The music bowl faded into limbo, and the Monroe Street parking lot remained above ground, open and ugly.

At the same time they were fighting the music bowl, the park defenders marched into another battle. The State Legislature passed an act permitting the Park District to transfer to the state the submerged land east of the United States Steel Company's South Works plant from 79th to 85th Streets. The land could be sold all the way to the Indiana border, the Legislature said—about 194 acres.

But Calvin Sawyier, the attorney who had fought McCormick Place in the courts, filed suit to stop the land transfer. He represented a concerned taxpayer, insurance broker Albert C. Droste. Sawyier and Droste contended the Legislature had no right to pass the act because the submerged land was held in trust for the people of Illinois, not for a steel company. The sale would set a dangerous precedent, they warned, for the misuse of other lakefront land.

They had another worry. Indiana had no landfill restrictions, so it was possible that the steel company could eventually extend its new plant farther into the lake beyond the state border, jeopardizing Rainbow Park to the north and Calumet Park to the south.

Sawyier and Droste lost in the local courts, but the State Supreme Court reversed that decision. It looked as though the plant expansion was stopped. The victory astonished the park defenders, who in recent years had not had much success in the courts.

They were right to be surprised. Several months later the court met again and reversed itself, upholding the local court.

When it became clear the sale could not be blocked, Sawyier urged the park commissioners to keep a strip of submerged land along the state border and turn it into a park. Thus, there would be a continuous park strip from Rainbow across the huge steel plant.

The park commissioners liked the idea. Then the 1970 business slowdown hit and U.S. Steel did not press for the purchase of the submerged land. The question of how far the new plant would stretch into the lake, and how it would relate to the parks, was temporarily averted.

Sawyier went into court again a few years after the U. S. Steel case to fight plans for public schools in the parks. The Park Commissioners had given the Chicago Board of Education permission to build in three parks, including the green meadows of Olmsted's Washington Park.

Again, Sawyier sued on behalf of prominent Chicagoans who

objected as taxpayers to the loss of their parkland. This time the *Chicago Tribune*, formerly on the other side of the park battles, supported the cause. A number of major civic groups filed friend-of-the-court briefs backing Sawyier. But the State Supreme Court refused to block the school construction.

The wording of the decision convinced Sawyier and leaders of the civic groups who supported him that the case actually was a long-range victory.

The court ruled that park property is held in "public trust" and that citizens have a right to go to court to protect that trust. For the first time, the court spelled out clearcut conditions that must be met before any change can be made in park property: A public body must control the specified area. The area must be devoted to public purposes and must be open to the public. The effect on the original use must be relatively small. None of the original public uses can be destroyed or greatly impaired. The inconvenience of losing the former use must be negligible when compared to the greater convenience of the new use.

The plans for the schools met these strict standards, according to the court. Would McCormick Place have met them? The groups that backed Sawyier doubted it and were optimistic. Sawyier himself called it "a substantial victory for those members of the public who wished to prevent the misuse of public trust property . . . . I have been involved in several past cases seeking to protect public land from misuse, and this is the first time I have ever voiced optimism over a decision."

By the mid-1960s, the park groups were fighting on a number of fronts. In summer, 1965, the city unveiled the $6,000,000 project to widen and straighten South Lake Shore Drive in Jackson Park. The men, women and children who fought for the park and faced the chain saws did halt the destruction, but the park was badly scarred. Acres of lawns vanished, and about 800 trees were destroyed. Part of the lagoon was filled in. The dense willows and oaks that lined it were slaughtered. The boathouse built during the Columbian Exposition was torn down.

That fiasco, and the bad publicity it gave city highway planners, apparently did not have much impact in City Hall. A few weeks after the chain saws halted in Jackson Park, the city announced a $6,500,000 project to widen and straighten North Lake Shore Drive, which meant chopping trees and tearing up lawns in Lincoln Park. The plan was revealed only a week before bidding was scheduled to open on the work, prompting the Lincoln Park Conservation Association to denounce the city's "secrecy."

The city's actions "indicate a lack of interest in public participation in public affairs that bodes ill for democratic government," said William Friedlander, head of the group. The quality of city planning "continues to be appalling. If only somebody would start thinking, instead of just pouring cement."

Both the *Chicago Sun-Times* and the *Chicago Daily News* had fought the city's Jackson Park plan and jumped quickly into the Lincoln Park fight. Again, women and children tied white bands around doomed trees. Again, city workers tore them down. Children in nearby schools stayed after class to make hundreds of new ones.

The protesters included some very big names. There was Mrs. Adlai E. Stevenson III, wife of the rising young Democratic politician and future senator; Mrs. Robert Merriam, whose husband had run for mayor against Daley and whose father-in-law had led the campaign decades before to depress the Illinois Central Railroad tracks and electrify its trains; Mrs. Lester Fisher, wife of the director of Lincoln Park Zoo, and Mrs. Bernard Rogers, wife of the president of the zoo.

At least one city official, architect Paul Gebhardt, Jr., denounced the plan. He emphasized that he was speaking as an individual, not in his capacity as a city architect. And as an individual, he was fed up with the pre-eminence of roads over open space.

"We have a beautiful lakefront," Gebhardt told *Daily News* urban affairs reporter Paul Gapp. "It's famous the world over. It is unfortunate to have the choicest part of Chicago taken away and used for people to speed through. I hate to see the park taken over as a traffic thoroughfare. There is already a concrete curtain preventing the people from enjoying their lakefront."

The section of Lake Shore Drive the city wanted to widen and straighten was not dangerous, he said—just crowded. "But you never satisfy the appetite of vehicular traffic in an urban setting."

The mayor, however, hinted that big parks such as Jackson and Lincoln were obsolete and therefore fair game for roads and other public works. "The thinking now is to have more small parks out in the neighborhoods rather than these large parks," he said. "The school-park complex is the thing that urban renewal experts are thinking about."

Yet the storm of citizen protest and the newspaper opposition worried him. One fall Sunday afternoon he took an unannounced stroll through the south section of the park, the part slated for demolition. Later that week he startled everyone by retreating from his earlier position. The road project would be sharply curtailed, he said, sparing the park and several hundred trees. "There's no more beautiful place to walk," the mayor said.

(But he bristled when a reporter suggested he had "given in" to the protesters. He had done no such thing, he said, and would never "give in" to any protesting group. "We still have our road improvement," the mayor said.)

There was not much time for rejoicing. The Park District had approved plans for a mall in the center of Lincoln Park Zoo that was a static, rigid crisscross of concrete walks divided by planters in concrete bins. It had been designed by the Park District's own staff in a style described by *Sun-Times* writer Ruth Moore as "early shopping center."

For an entrance to the zoo, the Park District planned to build heavy concrete bastion walls and large iron gates. All that heavy concrete not only clashed with the charming English-park concept the nineteenth-century landscapers had incorporated into the zoo—it totally overpowered it. The Chicago Chapter of the American Institute of Architects and the Lincoln Park

Zoological Society demanded that the Park District commission a high-calibre architectural firm to prepare new plans; the commissioners reluctantly agreed. The result was a beautiful free-form mall designed by Harry Weese and Associates. There were curving, tree-planted walks past a seal pool, past animals moving freely in natural rock-like terraces surrounded by moats instead of walls or bars. Plantings were used to conceal the walls of the animal houses. The Weese plan had an aura of strolling through a jungle or a seacoast and seeing the animals in their natural setting.

Harry Weese and his brother Ben had also included a step-by-step master plan for the zoo and its surrounding parkland. "It could be done piecemeal, as the money becomes available, rather than waste zillions of dollars on haphazardly planned projects," Ben Weese said. Most important, the plan preserved the sense of space and beautiful vistas—the views of meadows and water from tree-lined walks—created by the original planners.

The plan was architecturally sound, esthetically pleasing and practical. But that didn't mean city officials would accept it. To do so would be giving in too easily, setting a dangerous precedent. As in the case of Jackson Park and the Johnson, Johnson and Roy plan, a consulting firm that frequently got city business was hired to study the Weese plan. Incredibly, the consultant recommended the shopping-center Park District proposal. Contracts were let and work was begun.

But the protests were so strong, and from such prestigious sources, that work was halted soon after it started. Again, consultants were hired. But this time they were two respected landscape architects, Robert Zion of New York and Franz I. Lipp of Chicago. They prepared a "compromise" plan that was much more Weese than Park District.

Finally, early in 1971, work began on the new mall. Five years had been lost in the wrangling, and thousands of dollars spent in consultants' fees. But at least an inferior project had been blocked.

It was obvious, though, that this was hardly the best way to develop parks: the city produces an inadequate, low-calibre design; citizen and professional groups fight it; a new study is ordered; a consultant is hired to study the study; the fighting drags on for months or even years; finally, some sort of compromise is worked out.

Yet the badly needed master plan, promised first for 1968 and for certain in 1969 and definitely by 1970 and absolutely by 1971, still had not appeared.

Travesties of design and development popped up, sometimes too quickly for the park groups to stop them. In Lincoln Park, wrote Ruth Moore in the *Sun-Times*, "graceful old bridges, shelters and recreational buildings have been replaced by graceless new ones . . . a heavy expressway-type concrete bridge replaced an elegant old iron bridge."

Buff-colored heavy stone wings were added to the soaring glass delicacy of the old conservatory. "From a short distance," wrote Miss Moore, "they look like two buses parked there."

When more lights were needed, the graceful old ones were replaced by standard mercury vapor fixtures "more suitable to a parking lot."

Maintenance in Lincoln Park, as in Jackson Park, deteriorated. "One of the lagoons is so filled with algae that rowing through it is like rowing through a salad," Miss Moore wrote. The sandy soil of the dunes had eroded and was not repaired, exposing tree roots, slowly killing the trees and destroying the ancient natural landscape.

At least one Lincoln Park planning disaster was stopped by the park vigilantes. "We caught the Park District trying to build a parking lot right at the end of that beautiful lagoon in Lincoln Park that ends in a keyhole shape," recalled architect Ben Weese. Park officials dropped the plan after vigorous opposition from the community.

In addition to their day-to-day watchdog job, the groups organized to protect the parks had grave long-range worries. Early in 1967 the United States Supreme Court cleared the way for the Illinois Central Railroad to lease the air over its tracks and right-of-way. Immediately, the strip from the Chicago River south along the lakefront to 47th Street became one of the most sought-after pieces of real estate in the nation.

Urban planners warned that a wall of high rises might be built there, cutting the city from its lakefront unless Chicago officials exerted stringent controls over the development.

But the plans approved by the city for the first phases were not reassuring. The 83-acre tract bounded by Randolph Street, the river, Michigan Avenue and the lake—formerly a tangle of tracks, freight yards, unsightly billboards and warehouses—seemed destined to be the most people-packed real estate in the nation.

Illinois Center Corporation, a firm formed by the railroad, is developing the 83-acre site with Jupiter Corporation and Metropolitan Structures, Inc. They have received permission from the Chicago Plan Commission and from the city to construct apartment and office buildings that will range from 30 to 80 stories high. These will accommodate 35,000 to 50,000 residents, with a day-time working population of 80,000—a density that makes the distinguished urban planner Lewis Mumford shudder. "It will interfere with the magnificent sweep of Michigan Avenue," he warned. "There is no reason for grabbing the lakefront. There is plenty of decayed property near the downtown area that could be rehabilitated instead."

The city did not ask for ceilings on rent, prompting Mumford to predict that the new lakefront city would be "for the chosen few."

The Metropolitan Housing and Planning Council urged that any landfill connected with the new air rights development conform to the symmetrical design of Daniel Burnham's plan for the downtown harbor area. The Burnham peninsula at Roosevelt Road (12th Street) had been built, with its museum complex. Now the Illinois Central development could complete the plan with the northern peninsula.

Instead, the plans approved by the city suggested creation of a huge mass of landfill with no particular shape or relationship to the southern peninsula—and no relationship to Daniel Burnham's lakefront plan. But this was merely a suggestion on the city's part, a gesture to appease the conservationists; it was not a commitment. Whether there actually would be landfill, and whether it would conform to the Burnham plan, were unanswered questions.

The proposal also provided for a mammoth traffic interchange to accommodate the tens of thousands of cars that would move in and out of the high-rise city every day. Lake Shore Drive would be straightened and moved east of the development, with a towering cloverleaf heading west, north and south. The drive through Grant Park would be widened. The result, according to architect and city planner Harry Weese, would be "a great amorphous lump—a spaghetti of highways."

East of the "great amorphous lump" would be a lakefront park, but probably no more than 10 acres, hardly enough recreational space for that densely populated area.

When the plan came up for approval before the City Council, several independent aldermen argued against it. Alderman William Singer noted that developers originally had talked about 140 acres of parkland; now there were only 9.86 acres in the plans. What happened to the rest? Lewis Hill, head of the City Department of Planning and Development, told him not to worry. In time there would be more landfill, he said, and maybe someday the Monroe Street parking lot in Grant Park would be decked over with park. In the end, only 6 of the 50 aldermen voted against the plan.

Phase Two of the air rights development is the high-rise hotel, residential and commercial complex west of McCormick Place. Eventually, the air rights will be leased to developers all the way south along Burnham Park.

Preventing a long line of high-rise villages and mammoth traffic interchanges will require enormous energy from civic groups and, equally important, some different attitudes in City Hall.

While developers drooled over what to do with the air rights strip along the shore, Mayor Daley and top city officials talked about something just as momentous in the lake itself: an airport, built in an 11,000-acre basin of drained lake bottom east of 55th Street. Circular dikes would hold back the water.

Architects, urban planners and ecologists were horrified. A lake airport, according to recent estimates, would cost $1.2 billion, three to four times as much as a land airport. Certainly there were more pressing demands on tax dollars.

The problems of water, air and noise pollution along the lakefront would be staggering. Motels and hotels and other travel-connected industries inevitably would rise as close to the shore as possible. Traffic congestion would force expansion of Lake Shore Drive and more multi-million-dollar concrete cloverleafs.

John H. Shaffer, head of the Federal Aviation Administration, said the proposed high-rise development on the Illinois Central air rights meant the airport would have to be at least 6½ miles offshore to be reasonably safe. Costly and ugly causeways cutting into the lake, or enormously expensive tunnels, would have to be built.

The most logical site for a new airport, according to some urban planners, was the open land near Joliet, 40 miles southwest of the city. It was cheap, it was accessible to expressways and the boom that would develop around it would provide jobs for a new urban complex.

But city officials wanted those jobs and that tax revenue for Chicago. "People have been telling me how lucky we are to have a lake where we can build an airport," said William Downes, city aviation commissioner.

Patrick O'Malley, a vending machine executive who was president of the Chicago Plan Commission and a close friend of the mayor, announced that he was all for a lake airport, thus ending any hope that the supposedly independent commission would block it.

O'Malley's announcement seemed a betrayal of everything

the Plan Commission stood for in its early Daniel Burnham days, but it was hardly a surprise. He had served as head of the mayor's citizens committee to recommend plans for a sports stadium and had insisted on the lakefront site. The site subcommittee preferred a location west of the old Soldier Field over the Illinois Central air rights, but O'Malley and the mayor persuaded three of the four subcommittee members to change their minds and opt for Burnham Park. "It was a matter of money," explained Alderman Claude W. B. Holman, one subcommittee member who made the switch. He said O'Malley told the group that $50,000,000 in land costs could be saved by locating on "free" park property. (That sum was considered a gross exaggeration; so was the prediction by the mayor that the stadium could subsist without tax subsidy.)

The lone dissenter on the site subcommittee was Marshall Field, publisher of the *Chicago Daily News* and the *Chicago Sun-Times*. His newspapers vowed to fight the stadium plans. A few days after the mayor's announcement, both ran hard-hitting editorials and cartoons blasting "the continuing willingness to turn the lakefront over parcel by parcel to developments that substitute concrete and asphalt for the trees and grass and beauty Chicago so desperately needs to conserve," as the *Daily News* put it. The *Sun-Times* added, "Once again we must ask our fellow citizens to rally round and do what must be done to save the lakefront from further commercial encroachment and despoliation."

And the citizens did rally round. In an amazing display of concern, they sent more than a thousand letters to the two newspapers in five days. Nearly all of them were violently opposed to the mayor's plan. It wasn't economics that angered them, said *Daily News* editor Daryle Feldmeir—although they did object to the use of real estate tax money to back the stadium bonds. He said they were fighting "this new threat to spread concrete over still more precious acres of lakefront that ought to be kept green and beautiful and saved for the enjoyment of all."

It was a heartening sign that the citizens of Chicago had learned the lesson of McCormick Place. They were not going to let it happen again to their lakefront.

Daley, totally unprepared for that kind of reaction, temporarily backed down. After angrily accusing the protesters of being "talkers, not doers," he offered to switch plans and remodel Soldier Field instead of building a new stadium. It was not a satisfactory compromise. Anyone who has driven past Soldier Field on a Sunday afternoon when the Chicago Bears are playing there can see why. The 41 acres of parking lots that deface the lakefront from 13th Street to 20th Street are packed solid, so solid that some fans sit trapped in their cars for an hour after the game, shivering and wishing the Bears were back in Wrigley Field, close to the elevated rapid transit station. Narrow Burnham Park, east of Lake Shore Drive, also is packed with cars. Bumpers push against the park's lone ornament, the majestic old Roman column. The lawns around the Field Museum are full of cars, as is every other bit of unoccupied park property in the two and a half miles from McCormick Place to Roosevelt Road.

When the game is over, any motorist who happens to be heading somewhere on Lake Shore Drive will be caught in a massive traffic jam. Even the monstrous double-deck highway interchange towering over McCormick Place can't cope with 50,000 or 60,000 people heading home from a football game. Will another ugly highway interchange be built? Will more parking lots spread across the lakefront? Stadium planners have already pointed out that Soldier Field has room for only 7,000 cars if they are properly parked. For an adequate ratio of fans-per-auto, they say, Soldier Field should have room for 17,000 cars.

Mayor Daley said the extension of the downtown subway system, with a line to McCormick Place, will relieve the traffic congestion and parking pressure around the stadium. But that $750,000,000 subway program has been tied up in court tests, and none of the two-thirds federal contribution has been allocated. When it does begin, construction will take seven to ten years.

Although the outcome of the sports stadium fight was hardly a total victory for the conservationists, it did show that the public could be aroused to protect the lakefront. That meant the mayor would have a rough time pushing through his airport in the lake.

The groundwork for the airport proposal had been carefully laid in an old Chicago City Hall tradition: The Department of Public Works commissioned several real estate and engineering consultants who get a great deal of city business to study the proposal. The consultants were asked to decide whether or not a lake airport was feasible—not whether it was the best of several possible sites. They all pronounced it feasible.

But except for City Hall, some downtown business interests and the mayor's cooperative commission chairmen, there was little enthusiasm for an airport in the lake.

Richard B. Ogilvie, the Republican governor, and the state's two senators, Republican Charles H. Percy and Democrat Adlai E. Stevenson III, all promised to fight an "aquaport." It seemed doubtful that the necessary federal money could be obtained without their consent. And, as long as the economic slowdown cut into the air travel business, the pressure for a new airport faded.

But it seemed certain that O'Hare Airport, busiest in the world, would not be able to handle the traffic indefinitely. And it was just as certain that the mayor was not about to give up his airport-in-the-lake. In 1971, during his campaign for his fifth term, he frequently mentioned the aquaport. "It will help Chicago remain a great city," he said. "There is no use putting an airport 50 or 60 miles from the city." Later that year his Cook County Democratic Delegation in the Illinois Legislature succeeded in killing a bill that would have banned a lake airport.

Another problem surfaced in the 1960s: The junk floating in the lake. Water pollution from the gigantic industrial complex around Calumet Harbor became a matter of growing civic concern.

The Illinois attorney general, William Scott, concentrated on forcing United States Steel Corporation to stop dumping wastes from its South Works plant on 79th Street. U. S. Steel had to be the target, because the other major plants of the Calumet district were across the border in Indiana.

During the long court battle, U. S. Steel violated a 1968 pollution control deadline set by state and federal officials in 1965, violated a state of Illinois pollution control deadline of September, 1969, and was found in contempt of court for polluting the lake with five illegal oil spills in 1969 and 1970.

Finally, early in 1971, the company accepted a deadline of October 31, 1975, for completing a pollution control program.

The volume of its waste waters discharged from the plant would be reduced from 100,000,000 gallons a day to 3,000,000 gallons a day. Cyanide, phenol and ammonia discharges would cease altogether. The program, expected to cost from $8,000,000 to $12,000,000, could be a model for the nation, according to State Attorney General Scott. But would Indiana officials press as hard on the plants across the border? It took Illinois six years to win a firm commitment of compliance from U. S. Steel—and another four years was granted to complete the compliance. If a ten-year delay became established procedure, the outlook for the lake was grave.

Lake Michigan is considered too big to die completely from oxygen depletion, but in areas where pollution is heavy, such as the Calumet district, depletion could choke fish and plants.

Dozens of "outfalls" (ducts or pipes that empty the wastes from industry) pour into the shipping canal and the Calumet River and eventually flow into the lake. Indiana officials have not tried very hard to stop them. The U. S. Attorney's office in the Chicago-Northern Indiana district has tried, but one assistant U. S. attorney complained that the federal laws are "paper-toothed, almost impossible to enforce. They are no more than inspirational in nature. They merely inspire the states to eliminate pollution and espouse standards. Terms like persuade, conciliate, negotiate—it can take years to get over all the persuasion and negotiation and finally get an injunction."

Some scientists predict that by 1980 the Chicago shore from Milwaukee to Hammond will be a dead sea if contamination continues as its present rate. Drinking water will stink. Beaches will be closed. The lakefront parks, however gloriously they may have developed by then, will border a cesspool overrun with swampy growths.

The Metropolitan Sanitary District provided a hopeful note by proceeding with plans for a $650,000,000, 95-mile "deep tunnel" system. A storage complex 100 to 300 feet below ground would hold polluted water after heavy rains. Later, when the pressure subsided, the water would be pumped into existing sewage treatment plants. In the present system, heavy rains force untreated sewage to be discharged directly into the river and sanitary canals about 50 times a year. Four times in the last 15 years untreated sewage has been discharged into the lake to prevent the river from flooding after heavy storms.

Another development at the close of the decade was promising. Mayor Daley replaced three of the elderly park commissioners, those men who couldn't say no to roads and parking lots and missile sites and filtration plants and a convention hall.

All three of his replacements were departures from tradition. Two were young and interested in athletics and one had been genuinely concerned about parks for a long time.

Gale Sayers, the outstanding and overwhelmingly popular running back of the Chicago Bears, was named commissioner when he was 27, at the peak of his professional football career. He didn't have much time for Park District business.

Another newcomer, Franklin B. Schmick, a retired investment counselor, took on his new job with relish and some astounding statements. Schmick was particularly interested in Lincoln Park Zoo and had participated in some animal-finding expeditions. He knew something about Park District operations. Shortly after his appointment in 1969, he announced he didn't want loafing payrollers around and intended to do his best to get them to work or get rid of them. He also announced his opposition to private clubs with exclusive rights to portions of the parks, maintaining that "all parks should be available to all the people."

Getting rid of loafers would cut into the city's patronage system and deeply offend City Hall—something for which the Park District commissioners had little stomach. But Schmick's second complaint brought action. The commissioners ordered private clubs, such as the Lincoln Park Gun Club, to post big signs announcing that they were open to the public. Rates for public use had to be "reasonable," and approved by the commissioners. The gun club, for example, charges $2.40 for 25 shells—comparable to public skeet-and-trap shooting facilities. If any private clubs refuse to serve the public, the Park District superintendent is authorized to cancel their leases.

The third and most important appointment was Daniel Shannon, 35 when he assumed the presidency of the Park Board in May, 1969. He was an All-American football star from Notre Dame University and an executive in his father Peter's accounting firm. The firm did considerable business for the city and Peter Shannon was an old pal of Daley.

Shannon brought an accountant's orderly mind to the Park Board, and saw things he didn't like. "I find it inconceivable that an organization with . . . an overall budget in excess of $80,000,000 has not embarked on a major computer program," he said soon after his election by the board as president. There were no cost studies on equipment, he said; no studies analyzing travel time or work time for various jobs.

He talked about forming a new kind of mobile park, recreational units with theaters and carnival equipment that would travel into the neighborhoods. He wanted to enter Park District teams in tournaments, maybe even the Olympics.

But these were hardly the major issues facing him. The immediate future would require much more of Daniel Shannon. He would be called on to prove his courage and effectiveness by fighting City Hall, real estate interests, the mediocrity of the

Park District organization and the galloping demands of the automobile and the airplane. There were tough decisions ahead, and he showed little inclination to make them. The football stadium on the lakefront? He was all for it. An airport-in-the-lake? He could see reasons, he said, why it would be good for the city.

Those were not the answers the people of Chicago needed to safeguard their priceless lakefront.

Chapter 11

The Future

On Sunday afternoon I reached a lake surrounded by green lawns; in the sunlight the waters reminded me of silk and flashing diamonds; sailing boats were moving over them. It was the luxury of the Cote d'Azur . . . . There was nothing to remind one of the squalor with its human wreckage. Crowds of people were sitting on the grass in the bright sunshine; there were well-behaved loving couples . . . young people had got up a game of baseball on the lawns; and children were darting in and out of bushes playing at Indians . . . . A spring day made for leisure. Seagulls flew over the lake and Chicago appeared as a city huge, wealthy and gay."

Simone de Beauvoir,
after a visit to Chicago.

A lake surrounded by green lawns . . . . Crowds of people sitting on the grass in the bright sunshine . . . . Is there any other city in the world that has given its citizens, and its visitors, a treat like that for mile after dazzling mile?

It was Grant Park that charmed Simone de Beauvoir, French author-philosopher. But she would have been equally delighted with a dozen or more spots along Chicago's spectacular 30 miles of lakefront.

There are few green acres as pleasant as the fragrant Grandmother's Garden in Lincoln Park, nearly 80 years old but alive and healthy from early spring to late fall with old-fashioned plume poppy, meadowrue, bright orange Mexican torch, yellow-flowered yarrow.

Ride north along the bicycle paths for a mile and a half, past lagoons and grass, and see red and yellow and blue sailboats glide into the lake from Belmont Harbor, that carefully planned oval created in the 1920s with 290 acres of landfill, providing both a protected marina and a magnificent stretch of beach.

On the lawns west of the harbor, women spread their tablecloths and unpack fried chicken and corn-on-the-cob wrapped in foil to keep it warm, and all around are the joyful noises of summer: a child shrieks, a bird calls, a dog yelps.

A couple on a bicycle-built-for-two cruises down the Lincoln Park cycle path on their way to the dark and green and mysterious bird sanctuary, thick with tall marsh grasses, bushes and trees.

On narrow Burnham Park, across from the tightly packed, sagging old tenements and new high-rise developments of the Near South Side, children whoop down the high, wild-twisting slides of a psychedelic playground. Ribs smoke above barbecue pits, a perspiring Cub Scout troop marches doggedly on its way to a merit badge in outdoorsmanship, a puffing man in a sweatsuit jogs along the bicycle path and the lake is so full of white sails it looks as though flocks of great birds are swaying in the waves.

Jackson Park is marred by the wide unfinished highway cutting into its middle, yet you are still conscious of the water around you: the willows bending over the lagoon, the small harbors and rustic bridges planned 100 years ago. Trees are mirrored in still waters, couples with long hair flying in the shore breeze saunter over the paths, toddlers with ice-cream-smeared faces are wheeled in their strollers and a small black poodle chases a gigantic Irish wolfhound. From the beach across the highway, the pounding rhythms of a bongo band remind you that Indian music once throbbed along this lakefront, that for centuries families have cooked their meals here, lovers have strolled here, children have romped here.

Other cities have lakeshores that were free and open playgrounds in ancient times. But of them all, only Chicago preserved such a magnificent stretch for its people. Instead of warehouses and shipping docks and granaries and oil storage tanks, we have sand beaches, green lawns, beds of flowers and bicycle paths. Nearly 24 of the 30 miles of Chicago lakefront are publicly owned. More than half of the 2,800 acres in the seven shoreline parks is landfill, much of it paid for with tax dollars during hard times early in this century, when the city government had meager resources but a firm commitment to give the people open space and clear, clean lake vistas.

Since the days in the 1860s when the old city cemetery on the North Side was moved and Lincoln Park was created, about $160 million has been spent developing the lakefront park system, more than $57 thousand an acre. That is expensive, but Chicago's next expressway, the Crosstown, will cost a whopping $1 billion, not counting the homes and stores that will have to be torn up to make way for it.

The tax money invested in Chicago's lakefront parks has been more than equaled by citizens' efforts to preserve, enhance and expand them, from the early settlers who met in 1835 to demand "public ground" on the old Fort Dearborn site to the small children who lettered "Save This Tree" signs when they could have been outside playing ball.

Generations worked for future generations so that their grandchildren's children could know the freedom that comes from gazing at a green-and-blue, white-capped, restless, endless lake.

But the citizens also failed. In the last 50 years there have been disastrous developments that severely damaged the park system. Every one of them was the result of government action. It wasn't big business and cold commerce that robbed the people of their land, but the people's government, acting in the name of expediency.

It was government action that paved the lawns into highways, built a convention hall, an airport and two filtration plants on the shore, leased park land to the Army and to private clubs, grew slipshod in park maintenance, pushed for a sports stadium on the lakefront and an airport in the waters. Between one stretch of lakefront, from 13th Street to 23rd Street, the paving of the parks has been complete; nearly all grass has disappeared under block after block of parking lots on either side of Lake Shore Drive. Soldier Field and its lots fill the area west of the drive; McCormick Place and its lots, and the Meigs Field lots, fill the area to the east.

The tragedy of the lakefront story is that each of these governmental moves could have been defeated by an alert, organized, tough-minded public. Citizen action could have stopped the paving of the lakefront that began in the late 1920s and it could have stopped each of the inroads since then. When citizens did organize and protest in strength, they succeeded. They forced city officials to change their plans for an Oak Street beach overpass. They forced the Park District to drop plans for a huge music bowl on the Grant Park lawns. The impact of public opinion was clearly apparent when Mayor Daley announced plans on July 1, 1971, for a football stadium at Burnham Harbor. Then on July 8, after seven days of outraged citizen protest, he offered to "study" the possibility of turning Soldier Field into a football stadium instead.

That kind of forceful citizen action will be needed in the coming years to save the lakefront from a number of potentially catastrophic projects. There are four major problems:

The airport in the lake is still very much a threat. Early in 1971, Representative Abner Mikva, a Democrat from the Hyde Park neighborhood, and Representative Harold Collier, a

Republican from suburban Oak Park, successfully lobbied to kill a $500,000 airport-in-the-lake feasibility study from the budget of the National Aeronautics and Space Administration. The item was inserted in the NASA budget largely through the insistence of Representative John W. Wydler of New York, who wanted an airport in Long Island Sound. It was backed by most Chicago area Democrats in the House. Two of them, Roman C. Pucinski and John Kluczynski, vowed they would eventually get a feasibility study. And in City Hall, Mayor Daley and his top aides reaffirmed that they wanted that aquaport—even if it meant waiting several years for the proper time to push for it. If and when they do, they are certain they will have Park District President Daniel Shannon on their side.

Even if the problems of water and air pollution presented by a lake airport could be solved, and that is extremely doubtful, one by-product could not be tolerated: The beauty and tranquillity of the shoreline from 47th Street to 79th Street would be destroyed. A jungle of hotels, motels, gaudy restaurants, parking lots and highway interchanges would rise there, duplicating the ugliness around Chicago's O'Hare Airport.

Paving the lakefront with a super expressway is just beginning, as far as city traffic engineers are concerned—even though the eight-lane Lake Shore Drive already has created a formidable barrier between the people and the shore. John N. LaPlante, city traffic engineer, described his fond hope at an annual meeting of the Institute of Traffic Engineers: Relocate a six-block stretch of North Lake Shore Drive 300 feet into the lake to make it easier for motorists to travel between the drive and the Kennedy Expressway leading northwest. He said in a July, 1971, speech:

One expressway development that will have a very definite impact on the central area will be the upgrading and realigning of Lake Shore Drive. South of 23rd Street and north of Oak Street, Lake Shore Drive has been upgraded to an eight-lane parkway with grade-separated interchanges. Within the central area, this upgrading has not occurred, and the result is daily traffic jams . . . .

Between Ohio and Oak Streets, it appears that the only way to provide adequate interchanges for the Near North area will be to relocate Lake Shore Drive approximately 300 feet out into Lake Michigan.

When his speech was publicized, city planning officials promptly said it was only an idea; there was nothing specific, no plan. Yet it provided a chilling insight into the way traffic engineers regard the lakefront. Veterans of lakefront battles were convinced it well may be a long-range goal. As part of the "upgrading" of the drive in the central area, they expect city engineers to push for additional lanes through Grant Park, uprooting the formal gardens and lawns.

Daniel Shannon has accepted the city traffic engineers' scheme for realigning the drive in the area of the Illinois Center project, between Randolph Street and the river. "Lake Shore Drive would run east of the new buildings," he said, "with a cloverleaf heading west and north-south . . . ." It would be, as he described it, similar to the monstrous double-deck interchange towering over Lake Shore Drive at McCormick Place, blocking trees and grass and sand and blue water.

The city Public Works Department also has talked about a broad new Lake Shore Drive extension north of Hollywood Avenue, where it empties into congested Sheridan Road. This could be done by landfill adjacent to the existing shore or by a causeway in the lake that would form a lagoon between the new road and the existing shore. Either way it is essential that conservationists and expert landscape architects play key roles in the planning. Any landfill or causeway should be designed to enhance and expand lakefront parks on the Far North Side, which badly needs more recreation space.

The third major threat is the development over the Illinois Central Railroad right-of-way. The Illinois Center project that Shannon mentioned is the first phase, rising rapidly with the

$100,000,000 international headquarters of the Standard Oil Company (Indiana). This 83-acre phase is expected to be completed by 1985. Work began on the second phase, a huge strip of 160 acres from 11th Place to 31st Street, with the construction of McCormick Inn, west of McCormick Place. This strip also will be a commercial-residential complex, according to IC Industries, an offshoot of the railroad and parent to a number of newly formed real estate development corporations. IC Industries has picked a name for it, Chicago Lakefront South, and is working on plans with two other firms, Ogden Development Corporation and Charles Luckman Associates. If the density of Chicago Lakefront South approaches that planned for Illinois Center, the effect on the shore—and the people living west of the railroad area—could be catastrophic. A two-and-a-half-mile strip of densely packed high-rises could be built there, cutting off the rest of the city from the lake.

A fourth threat to the lakefront is the continuing fouling of the waters. This has been most serious in the Calumet Harbor-Indiana Harbor industrial belt to the south, where industry dumps its wastes into the lake and into streams that feed the lake, and in Chicago's North Shore suburbs, where raw and inadequately treated sewage is discharged into the lake. For several years swimming has been banned or limited at many beaches from the suburb of Highland Park north to the state line.

United States Steel Corporation's South Works, the giant plant located within Chicago city limits, is under court order to recycle all its waste water now discharged into the lake by October 31, 1975. Company officials say they intend to keep the deadline—unless there are serious and prolonged work stoppages in supplying plants or in the steel industry itself. Strikes always are a threat, so that deadline, imposed in 1971 after a six-year legal fight, is tenuous. But the other giant plants, south of the Chicago border in Indiana, haven't even reached the first step of agreeing to a recycling deadline.

The Indiana Stream Pollution Control Board recently published data showing that Inland Steel Company and Youngstown Sheet and Tube Company in East Chicago and the United States Steel Company plant in Gary annually pour thousands of tons of iron, oil, and nitrates into Lake Michigan or the Indiana Harbor Canal, which flows into Lake Michigan.

The fouling of the northern waters also is likely to continue for several more damaging years. The North Shore Sanitary District has plans for a massive expansion program to replace its obsolete plants in Highland Park, but endless lawsuits and administrative battles have delayed it. A $35,000,000 bond issue for the program was approved in 1968, but residents of the area who were dissatisfied with the plans managed to block them for more than two years, during which the estimated cost rose from $35,000,000 to $85,000,000. It will be mid-1973 by the time additional funds are obtained and the project finally is completed. And, when the Illinois League of Women Voters complained that the program would not totally eradicate the fouling of the lake, the Illinois Pollution Control Board admitted that the League was correct: The program would permit some lake damage. But, said the board, there is "no practical alternative." Better equipment is not on the designing boards and would be prohibitively expensive if it did exist.

To meet these grave threats to the lakefront there must be, above all, a renewal of the cooperative spirit between the people and their government that created the original parks, the spirit that set the precedent for an open, clean shoreline. And, before any specific goals or plans are set, there must be a recommitment from government to that principle of an open, clean shoreline.

The need for such a specific commitment was recognized by the Northeastern Illinois Planning Commission, a group created by the Illinois Legislature and made up of appointees of the governor, the mayor of Chicago, and the board presidents of the six counties of the region. The commission spent two years preparing a detailed open-space plan for the six-county region to meet the federal requirement that a comprehensive plan must be drawn before local governments can qualify for federal funds to acquire and develop open-space areas. The plan was completed in April, 1971. Its statements concerning the Lake Michigan shore are clear and succinct:

The presence of Lake Michigan is an asset of incalculable environmental and recreational value to the residents of Northeastern Illinois. The major lakefront park system immediately adjacent to some of the most intensively urbanized portions of the metropolitan area already constitutes a development unique in the nation and, perhaps, the world. However, boating on Lake Michigan is largely restricted to those with access to larger boats because there is a very limited sheltered area which is necessary for safe handling of small craft. The deficiency of mooring and docking facilities further reduces the boating potential of the lake. Boat rental facilities, now almost non-existent, could greatly increase the level of use of Lake Michigan and its harbor areas.

In the past, limited landfill development has been used for the purpose of providing recreational opportunities along the Lake Michigan shoreline. These efforts must be intensified in order to utilize the full recreational potential of Lake Michigan. The construction of peninsulas and islands in Lake Michigan may be the only feasible method of providing the vast areas of regional open space needed by the citizens of Chicago and adjacent highly urbanized communities. Construction of these peninsulas and islands must follow extensive studies and legislation to safeguard the ecology of the lake and protect its shoreline from encroachment by incompatible land uses.

There are several sites along the Lake Michigan shoreline which may become available for redevelopment. These include governmental properties and private golf courses which could be sold and developed for non-open space uses. Immediate contacts with the present owners of these sites may insure their preservation and future development as regional open space.

The properties referred to by the commission include two of

major importance: South Shore Country Club, occupying a choice spot from 67th Street to 71st Street in Chicago, and Fort Sheridan, the Army base north of Chicago that covers two miles of lakefront in Lake County. The South Shore Country Club, an all-white relic in a neighborhood that has become largely black, is on the verge of folding. Its magnificent facilities would be an enormous boon to the crowded South Side, with its densely populated low-income ghettoes. The U. S. Defense Department has transferred the 5th Army from Fort Sheridan to Fort Sam Houston, in Texas, leaving only a small contingent on the base. These 1,400 troops easily could be quartered in the west section of the reservation and a lakefront park could be developed on the beautiful bluffs and sand beaches of the east section.

The commission made three specific recommendations regarding Lake Michigan in its open-space report:

1. As a long-range policy, rights should be acquired progressively along the entire Lake Michigan shoreline in Illinois to allow for continuous public access to and utilization of all areas which are safe for recreational purposes.
2. No further development of any kind along the Lake Michigan shoreline or into the waters of Lake Michigan should be permitted unless it can be demonstrated that such development will not have adverse effect on the ecology of Lake Michigan and adjacent areas.
3. Further landfill in Lake Michigan for open-space purposes is to be encouraged, but only if no adverse ecological effects are produced. Any further landfill for other than open-space use should be permitted only if it is compatible with the open-space use of this extremely valuable resource.

The formal adoption of these recommendations by the Chicago City Council, the Chicago Plan Commission and the Park District could be an excellent start in renewing the spirit that built the lakefront parks.

Another proposal, with similar but more stringent recommendations, was prepared by a state legislative study commission. It was presented early in 1971 to the Illinois Legislature by State Representative Robert E. Mann, a Democrat from Chicago's Hyde Park neighborhood.

The commission's "Lakefront Bill of Rights" included these prohibitions:

1. No pollutants shall be deposited, discharged or allowed to run off into Lake Michigan or any other waterbody within the state of Illinois from sources originating within the land adjoining Lake Michigan. All water-borne domestic sewage and industrial wastes must be recycled or—as a second priority—there must be confined disposal of all nonrecycled pollutants which threaten public health or safety or the maintenance of a beneficial ecological system.

2. The highest and best use of the land abutting on Lake Michigan in Illinois is declared to be for open space, park and recreational purposes. To this end, the maximum possible Lake Michigan frontage land shall be acquired in fee simple or lesser interest by the state of Illinois or by any municipal corporation within whose jurisdiction such land is located, and such land shall be held as open space and developed for public park and recreational purposes. The development or new use of land abutting on Lake Michigan after the effective date of the act for purposes incompatible or interfering with the open space use of the lakefront shall be prohibited.

3. Maximum physical and visual access to Lake Michigan shall be afforded to all citizens of Illinois and no land adjoining Lake Michigan shall, after the effective date of the act, be developed in such a way as to restrict public rights of access to Lake Michigan and the clear and full visibility of the lake to the general public.

4. There shall be no impoundments or fill of the waters of Lake Michigan for the purposes of landfill, causeway, polder construction or for any other purposes, except as authorized in the act. There shall be no impoundment or fill of the waters of

Lake Michigan without the authorization and approval of the legislative council or governing board and chief executive officer of the municipality or municipalities adjacent to the proposed impoundment or landfill site, the General Assembly of the State of Illinois and the Governor.

Thus, with one legislative stroke, Mann's bill would have wiped out most of the serious threats to the lakefront. It would have effectively outlawed an airport in the lake, placed strict controls on landfill, required open-vista easements in the Illinois Central right-of-way developments, possibly have outlawed a sports stadium on the lakefront (on the grounds that it would be "incompatible or interfering with the open space use of the lakefront") and protected the waters from death-by-pollution.

Perhaps its time had not yet come. The bill died in the 1971 spring Legislative session without coming to a vote. In the fall session vigorous lobbying by conservationist groups pushed it through the House of Representatives, but Chicago Democrats loyal to Mayor Daley kept it from coming to a vote in the Senate.

"This is another move to take the authority that belongs with the local government and place it in Springfield," the mayor said at a City Hall press conference. Besides, he added, it would "make it almost impossible to build a lake airport."

Supporters of the bill, including the Independent Voters of Illinois, the Illinois League of Women Voters, the Audubon Society, the Illinois Wildlife Federation, and the Citizens Action Program, born several years ago to fight pollution, prepared for a tough fight in 1972. State Senator Charles Chew, one of Daley's loyalists, warned: "We are all set to kill it."

The Bill of Rights and the recommendations of the Northeastern Illinois Planning Commission would accomplish the first, most important step in safeguarding the lakefront: Establishment of a firm framework for future development, with broad principles that would permanently outlaw any destructive ventures.

But within that framework, a more specific master plan is needed. This plan should set forth guidelines not only for major expansion projects, but also for small yet significant details such as design and placement of lights and landscaping to disguise the roads that cut up the parks.

Such a plan must be reviewed by neighborhood organizations and the civic and professional groups that have invested much time and energy in preserving the parks. These include the Metropolitan Housing and Planning Council, the City Club of Chicago, the Chicago Chapter of the American Institute of Architects, the Midwest Open Lands Association, the League of Women Voters, the Illinois Audubon Society, the Izaak Walton League, the Sierra Club, the Chicago Heritage Committee, the Federation for an Open Lakefront and the new and aggressive Businessmen for the Public Interest and Citizens Action Program.

In the past, the city government often neglected to seek community advice and consent to its public projects. There is pressure for citizen participation in police management and school administration and public health development, and there should also be citizen participation in the development of public parks and recreational facilities. It is, after all, the people's land and the people's shore.

A master plan also must include sound provisions for financing any proposed capital improvements. This, too, has been omitted from past City Hall and Park District lakefront proposals. The result has been predictable: The improvements never materialized.

With a good master plan approved by the community, more federal aid would be available. For several years Washington has demanded community participation and sound, long-range planning before granting funds to local governments. Chicago knows this only too well. It has lost millions in federal money for housing programs because of deficiencies in planning.

The Chicago Park District, which has been getting a paltry $1,000,000 to $2,000,000 a year in federal money toward its

$100,000,000-a-year budget, should receive far more from federal programs designed to expand recreation space in congested areas.

"I don't think Chicago has had the federal funds it ought to have," admits Daniel Shannon.

That's true. But first his board of commissioners and city officials must produce a good plan and proof of citizen participation in development of that plan.

The Johnson, Johnson and Roy study commissioned by the city for $100,000 in 1966 was to give specific guidelines for a master plan. But a year after the landscape architectural firm completed it, the study was still under wraps in City Hall. After the furor over the mayor's sports stadium proposal, City Planning Commissioner Lewis Hill hinted that the long-delayed report may never be released as originally prepared by its authors. The city would not make the study public, Hill said, "until we decide the thing does what we want it to do."

"It's really our decision, not theirs [Johnson, Johnson and Roy], about when the study is finished," Hill added. "If we want something more, they'd have to do more."

The Johnson, Johnson and Roy report emphasizes park and beach development and discourages any "major facilities attracting large surges of attendance on a regularly scheduled basis" on the lakefront, such as a sports stadium. It also runs counter to city policy on highway expansion, sharply criticizes the lack of pedestrian access to the lake and is cool to an airport in the lake. Unless city officials shift some of their thinking on the lakefront, it is likely they will be extremely reluctant to release the full report with all of its conclusions and recommendations. Citizen pressure, of course, could do much to force the release.

These are some goals that should be incorporated in a good master plan, based on the Johnson, Johnson and Roy study and the expertise of the civic leaders who have worked to preserve the lakefront. Some of these goals have been endorsed by Park District President Daniel Shannon. Another has the tentative approval of Lewis Hill, city planning commissioner. Some, particularly those pertaining to highway development and traffic control, are not acknowledged by city officials. Others, the ban on a lakefront sports arena and an airport in the lake, run counter to stated city objectives. Hopefully, though, there is enough agreement on these goals to lay a foundation for a solid master plan:

1. Adopt the principle that all of the 30-mile Chicago lakefront should be publicly owned.

Both the citizen groups and the new leadership of the Park District are committed to this principle. One major area still to be acquired is the two-and-a-half-mile strip from the northern boundary of Lincoln Park to the city limits, where individual property owners hold about a third of the shoreline, preventing extension of the park. The Park District has slowly been acquiring the shore under an old agreement that property owners may fill in the lake up to a certain point in exchange for surrendering shore rights beyond that line. As the old single-family mansions on the Far North Side are torn down to make way for high-rise apartment buildings, real estate developers have sought permission to build into the lake—and have turned over their riparian rights to the Park District. But this is an extremely slow process. In the past 35 years, only about a mile has been acquired.

Another major area not publicly owned is the South Shore Country Club, that spacious establishment that looks as if it were created for an F. Scott Fitzgerald novel. It has a fine beach and harbor, a nine-hole golf course, a riding stable, lawn bowling and trap-and-skeet shooting. As whites moved out of the South Shore area and blacks moved in, the club's membership dwindled. Real estate developers and Park District commissioners have watched its slow death like vultures, both eager to pounce on that magnificent stretch of shore.

"I'm dedicated to acquiring it, as soon as the club throws in the sponge," says Daniel Shannon. The Park District has the right of eminent domain and could condemn the land and force a sale if the club should fold, but Shannon points out it is extremely valuable property. Real estate developers could shoot the price much higher than the poverty-stricken Park District currently could afford. In that case, the club's directors could go to court to block Park District condemnation, charging that the commissioners are not offering fair market value.

It would be tragic if the Park District lost this prized recreational facility because of lack of funds. Another high-rise village would rise on the lakeshore. It is imperative that the Park District plan now to acquire the necessary money to buy the South Shore Country Club, exploring every possible source of funding.

The third major area not used for public recreation is the United States Steel Corporation's South Works plant and the adjoining Calumet Harbor district, from 79th Street to 95th Street, where Calumet Park begins.

The harbor undoubtedly will remain a harbor. But there is hope for a park in front of the steel plant. Shannon says he will approve the sale of the submerged land U. S. Steel wants for plant expansion only if the company agrees that the Park District can develop a recreational strip along the shoreline, as it has done along the shoreline of the new McCormick Place.

Shannon, however, says he will drop all plans for the sale of the submerged land if there is evidence the landfill would pollute the lake or upset lake ecology. "I'd prefer no landfill at all," he says.

2. Preserve and enhance the tranquillity of the parks.

The greatest single threat to the lakefront parks is the automobile, according to recreational planning authorities. They want to reverse the trend toward enlarging park roadways into high-speed thoroughfares. On weekends and holidays, they suggest slowing traffic on Lake Shore Drive through the parks or banning it altogether.

They deplore the massive concrete interchange that blights the lakefront west of McCormick Place and want a ban on any additional spaghetti-configurations along the lake, including the one proposed for the Illinois Center area. Good highway design, they say, should be able to produce alternatives.

All parking lots in the parks should be decked over with greenery; particularly the Monroe Street lot in Grant Park, the huge lots stretching for blocks south of Soldier Field and the lots that almost overpower the Museum of Science and Industry in Jackson Park.

In addition, the Illinois Central Railroad tracks that divide Grant Park from north to south should be covered. Decking them with lawn would make it possible for pedestrians to stroll from Michigan Avenue to the lake without going out of their way to find a bridge over the railroad.

The Johnson, Johnson and Roy study insists that Lake Shore Drive must not dominate the lakefront and divide its parks. The firm's final report recommends sensitive designs, not the stiff engineering layouts produced in the past by the city, to make the drive more scenic. The report also asks that city traffic engineers carefully calculate speeds to permit optimum traffic flow.

In Lincoln and Jackson Parks, the report recommends depressing the drive in several places and providing wide, tree-planted overpasses and landscaping along the embankments.

So far, the only one of these points specifically endorsed by the Park District and city officials is the decking of the Monroe Street parking lot. As the Illinois Center air rights development progresses, they say, the Park District will begin work on the parking lot.

3. Make the parks more accessible to the people.

This is a major proposal of the Johnson, Johnson and Roy report. The firm recommends bus lines and subway extensions terminating at the lakefront parks, and an emphasis on well-designed pedestrian overpasses to the parks.

4. Amend city zoning ordinances to set up a classification of land use for lakefront parks and future landfill extending into the lake.

There should be strict and explicit controls on construction within this classification. The Park District should not be permitted to transfer or lease land in this zoning classification to other public bodies for schools, filtration plants, convention halls, airports or other public works, as it has done in the past with such disastrous results.

5. *Establish a new zoning classification for "park frame" land, the property bordering the lakefront parks.*

This would give the city power to preserve open-ended vistas to the lake, and prevent a wall of tall buildings rising like a gigantic concrete-and-steel curtain infront of the shoreline parks.

This new zoning classification should include guidelines that would prevent inferior design in new construction near the lake.

An example of what happens without such special powers and watchfulness on the part of the city is the two-mile strip of high-rise apartments north of Lincoln Park along Sheridan Road. "They function as a physical and psychological barrier for all the communities to the west, seriously jeopardizing the public's access and view of the lake," said the Metropolitan Housing and Planning Council in a statement to city planning officials. What happened on Sheridan Road, the council said, must not happen again on the Illinois Central Railroad right-of-way developments.

6. *Prepare for I.C. developments on the South Side by establishing density patterns, preserving the existing "grid" street system, and providing scenic "see through" areas of open space leading to the lake.*

Chicago's traditional street pattern allocates heavy usage to streets at half-mile intervals, medium usage to quarter-mile streets and low-intensity usage to streets in between. In its recommendations to the city on I.C. guidelines, the Metropolitan Housing and Planning Council urged that this grid system be incorporated into the new developments. This would integrate the strip into the fabric of the city, and would insure open vistas to the lake.

The council's plan was prepared by a task force headed by Walter Netsch, a top architect and designer of the striking Chicago Circle Campus of the University of Illinois. He recommended that high-intensity development, both residential and commercial, be restricted to the areas adjacent to mile streets, with medium-intensity development at half-mile streets and low-intensity development, restricted to low-rise buildings, at quarter-mile streets. At two-block intervals there would be park strips, open vistas with easy access to the lakefront parks and beaches.

The council also urged that any concurrent developments in the area immediately west of the air rights should be "an integral part of the unified overall development plan for the lakefront land and subject to the same criteria as if it were legally defined as air rights."

This concept, developed by architects, city planners and civic leaders who are active in the council, would insure that the public could "see through" the new developments. Instead of a solid north-south wall blocking the lakefront, there would be a series of east-west "slots" every two blocks.

Lewis Hill, city planning commissioner, told the council leaders his department was thinking along the same lines—a hopeful sign for the future of the lakefront.

7. The professional staff of the Park District must be up-graded. It must engage high-calibre consultants, not just the City Hall favorites, and it should work to improve its relationships with the public.

A top-flight professional advisory committee, perhaps in the form of an Arts Commission, should pass on all park improvements and developments, insuring they are suitable in both design and function.

8. Navy Pier, growing old and obsolete, should be converted into a recreation area, with emphasis on marina development. There is room for an indoor swimming pool, gymnasiums, restaurants, handball and tennis courts and bowling alleys.

The city administration has plans along these lines for Navy

Pier. As part of a $5.2 billion public works program scheduled for 1972 through 1975, Navy Pier would be thoroughly renovated. The project, priced at $16,900,000, would include a $2,500,000 marina for pleasure craft and nearly that much for indoor tennis courts and a gymnasium. The pier's old auditorium would be renovated and a restaurant added. The eastern end, jutting into the lake, would be landscaped into a park. The pier would continue to be an overseas shipping terminal with port facilities. But this public works program is contingent upon federal and state aid that has not been granted, so there can be no precise timetable. The biggest item on the $5.2 billion public works agenda remains the $1 billion Crosstown Expressway, gobbling up nearly 20 percent of the four-year budget.

If Navy Pier does become a recreational center, care must be taken to prevent it from becoming cheap, gaudy and overly commercial. At one time Mayor Daley was enthusiastic about the proposal of a private company to build a $12,000,000, 920-foot "space needle" on the top of the pier, with a revolving restaurant on top. Fortunately, the Federal Aviation Administration said it would be too dangerous and the idea was killed.

9. Curtail to the absolute minimum all polluting influences on or near the lakefront: noise, air, water and visual blight.

10. Don't build an airport in the lake or a super sports stadium on the shore.

The ideal, of course, would be to raze Soldier Field, return the space to parkland and build a sports stadium on the fringe of the downtown area, where it could replace deteriorating structures and be convenient to public transportation and restaurants. And a second ideal goal would be to close Meigs Field, transfer its operations to Midway Airport, and convert Northerly Island to parkland, as Daniel Burnham and the civic leaders of the early 1900s had intended.

But to be practical, it is probably more worthwhile now to concentrate on preventing further destruction of the lakefront than to work for goals that may be impossible to achieve.

Lee M. Burkey, president of the Northeastern Illinois Planning Commmission, stated the dilemma well in his introduction to the Regional Open Space Plan: "Once lost, open space is exceedingly difficult, if not impossible, to reclaim. There can be no escaping the fact that now is the time—the only and last time—to prepare for future open-space needs."

Once there is a commitment to preserve the lakefront for the people, free and open, then there is no limit to the improvements and refinements that could be made in the park system. Chicago is blessed with a brilliant group of architects and planners who have an array of exciting proposals. C. F. Murphy Associates, the architectural firm that designed the vastly improved post-fire McCormick Place, has designed a music bowl that would cover the Monroe Street parking lot in Grant Park. Harry Weese, architect and city planner, has prepared plans for a one-way street system in Grant Park that he believes would reduce pressure for expanding Lake Shore Drive. He also has plans for public transportation lines feeding directly to the Buckingham Fountain area; at present, you have to be very quick on your feet, dodging traffic on Michigan Avenue and on the semi-circular Congress Plaza, to get to the fountain. He has sent his ideas to the Park District and City Hall for comment, but has received scant attention and no encouragement.

Boating is one area that definitely needs expansion. As the Northeastern Illinois Planning Commission said in its open-space report, boating has become a privilege of the affluent, who can afford their own crafts and who have enough clout to secure a mooring. There is such a critical shortage of mooring space, particularly in the central area, that many families who could afford a boat have been discouraged from buying one. And for those who can't afford one, boating on Lake Michigan is too farfetched even to dream about.

Why not create a protected harbor, with small boats for rent to families who can't buy their own? Chicago needs a harbor secure enough for little, inexpensive craft. An ideal spot would be west of the filtration plant near Navy Pier. Because of the ecological dangers, landfill must be used judiciously, but this would be a perfect use for it. By extending the filtration plant area to the north and east with landfill, a large wedge-shaped harbor would be formed between the fill and the beach. Here, families could rent dinghies for delightful holidays on the waters, secure and safe in a protected setting.

In the long rectangle of water between Navy Pier and the filtration plant, a series of floating docks could be constructed. The south side of the pier is used for commercial shipping, but this north side would be ideal for pleasure craft.

Another spot that needs landfill is the east end of the Illinois Center development. The 35,000 to 50,000 who will live here must have recreation space. The fill should conform to the configuration of the fill off Roosevelt Road, completing the symmetry Daniel Burnham incorporated into his beautiful Grant Park design. Bicycle paths on this new parkland would link North Side parks with the South Side.

The Johnson, Johnson and Roy plan calls for extensive landfill in the form of great, curving, eight-mile-long islands, three or four miles offshore. In between the islands and the shore there would be protected water for small boats. But, in the years since the plan was commissioned, there has been growing concern about the ecological effects of dumping fill in fresh water. It would be wiser—and certainly cheaper—to restrict fill to areas where it is acutely needed, such as the filtration plant-Illinois Center region.

A second area badly in need of additional park space is the Far North Side, from Hollywood Avenue, where Lincoln Park ends, to the city limits. This two-and-a-half-mile stretch has only one small lakefront park and a number of sandy street-end beaches. Because the Park District program of acquiring riparian rights in this strip must inevitably move exceedingly slow, architect-planner Harry Weese suggests an alternate plan. He would create a long peninsula of landfill, extending from Hollywood Beach north to the city limits. This would be a splendid park and beach area, and the lagoon between the peninsula and the shore would be ideal for small-craft rentals. There is no spot, at present, where a small boat is permitted to land on the Chicago shore; this new boating area could provide such a spot. Weese also envisions a few restaurants on the shore, "because I think it's nice at some points to have the city touch the lakefront, to have waves almost splash the windows of the dining room."

These are the dreams of the future. All of them, and more, could be realized if the people of Chicago and their government join forces, as they did in the late 1800s when Frederick Law Olmsted inspired them with his city-circled-by-parks, and again in the day of the Burnham Plan, when the vision of a beautiful park-along-the-lake spurred voters to approve mammoth public works bond issues.

Citizen pressure accomplished miracles in those days. Today, when the city is more affluent, better informed and just as desperately in need of open space, the people of Chicago could do it again.

There is one other goal for the future that has nothing to do with expansion or construction or design. But in a way it may be the most significant of all.

Children should be inspired to know and appreciate the lakefront through special courses in schools, special materials at public libraries and classes at Park District facilities.

The history of the lake should be taught to them, from the Ice Age to present-day struggles. Children would be fascinated by stories of the precarious existence of the maritime plants and animals living in their lake.

All of this would instill in Chicago-area youngsters a love of the lakefront that, sadly, may have faded in recent decades. Too many children grow up in the city and its suburbs without seeing the lake, without knowing that this city, alone among all the great cities of the world, has preserved most of its waterfront not for commerce and industry, not for the wealthy, but for the pleasure of all of its people.

They should be taught how to use the lake, and they will

value it. And, when their turn comes, they will fight to protect it.

"The lakefront by right belongs to the people," Daniel Burnham said. "It is a living thing, delighting man's eye and refreshing his spirit . . . It should be made so alluring that it will become the fixed habit of the people to seek its restful presence at every opportunity."

But it does take work to keep the lakefront free. There was such great effort in the past, and so much more still is needed. Yet for such a treasure, could anyone say the price is too great?

Bibliography

Books

Andreas, Alfred T. *History of Chicago.* 3 vols. Chicago: A. T. Andreas, 1884–86.

Angle, Paul M. *Prairie State: Impressions of Illinois, 1673–1967, by Travelers and Other Observers.* Chicago: University of Chicago Press, 1968.

Baker, Nina Brown. *Big Catalog.* New York: Harcourt, Brace and Co., 1956.

Banfield, Edward C. *Political Influence.* New York: The Free Press, 1961.

Burnham, Daniel H., and Bennett, Edward H. *Plan of Chicago.* Chicago: The Commercial Club, 1909.

City of Chicago Department of Development and Planning. *The Comprehensive Plan of Chicago.* Chicago: City of Chicago, 1966.

Cromie, Robert. *The Great Chicago Fire.* New York: McGraw-Hill, 1958.

Dedmon, Emmett. *Fabulous Chicago.* New York: Random House, 1953.

Halsey, Elizabeth. *Development of Public Recreation in Metropolitan Chicago*. Chicago: The Chicago Recreation Commission, 1940.

Hansen, Harry. *The Chicago. Rivers of America Series*. New York: Rinehart and Co., 1942.

Kirkland, Josephine and Caroline. *History of Chicago*. Chicago: Dibble, 1892.

Kogan, Herman, and Wendt, Lloyd. *Lords of the Levee: The Story of Bathhouse John and Hinky Dink*. Indianapolis: The Bobbs-Merrill Co., 1943.

Kogan, Herman, and Wendt, Lloyd. *Chicago: A Pictorial History*. New York: E. P. Dutton and Co., 1958.

Lewis, Lloyd, and Smith, Henry Justin. *Chicago, the History of Its Reputation.* New York: Harcourt, Brace and Co., 1929.

Pierce, Bessie Louise. *A History of Chicago*. 3 vols. New York: A. A. Knopf, 1937–57.

Quaife, Milo M. *Chicago and the Old Northwest, 1673–1835*. Chicago: University of Chicago Press, 1913.

Rex, Frederick. *The Mayors of the City of Chicago, 1837 to 1933*. Chicago: Municipal Reference Library, 1946.

Royko, Mike. *Boss: Richard J. Daley of Chicago*. New York: E. P. Dutton and Co., 1971.

Schroeder, Douglas. *The Issue of the Lakefront: An Historical-Critical Survey*. Chicago: Chicago Heritage Committee, 1963.

Wade, Richard C., and Mayer, Harold M. *Chicago: Growth of a Metropolis*. Chicago: University of Chicago Press, 1969.

Magazines

Clayton, John. "How They Tinkered with a River," *Chicago History, Magazine of the Chicago Historical Society*. Spring, 1970.

Clifton, James A. "Chicago Was Theirs," *Chicago History, Magazine of the Chicago Historical Society*. Spring, 1970.

Newman, M. W. "A Funny Thing Happened on the Way to Megalopolis," *Inland Architect*. December, 1970.

Newspapers

Chicago Democrat, November 4, 1835.
Chicago Daily Journal, July 13, 1847.
Chicago Tribune, April 1–30, 1857.
Chicago Tribune, February 4, 1868.
Chicago Tribune, March 19, 20, 28, 1873.
Chicago Journal, April 7, 1874.
Chicago Tribune, April 8, 1874.
Chicago Tribune, December 29, 1890.
Chicago Tribune, March 31, April 1, 1891.
Chicago Tribune, August 28, 1892.
Chicago Tribune, "Souvenir Map of the World's Columbian Exposition," 1893.
Chicago Daily News, March 19, 1948.
Chicago Daily News, March 12, 1956.
Chicago Daily News, July 2, 1956.
Chicago Daily News, August 6–8, 1956.
Chicago Daily News, May 4, 1957.
Chicago Daily News, May 23, 1957.
Chicago Tribune, November 13, 1960.
Chicago's American, February 21, 1966.
Chicago Daily News, 1962, to 1971, numerous articles on Chicago parks.
Chicago Sun-Times, 1962 to 1971, numerous articles on Chicago parks.
Chicago Tribune, 1962 to 1971, numerous articles on Chicago parks.

Documents

Beecher, W. J. "Chicago's Ancient Coral Reefs," *Science Notes*, Chicago Academy of Sciences, 1965.
Central and North Filtration Plant–Proposed Sites. Chicago: Chicago Plan Commission, 1949.
Chicago Harbor Commission Report. Chicago: Chicago Harbor Commission, 1909.

Index